The Turkic Peoples in Medieval Arabic Writings

Translating a collection of the most important descriptions of the Turks found in medieval Arabic texts into English, this book aims at delineating the coming of the Turkic peoples in the eleventh century, their military successes in Iran and Iraq, and the emergence of the sultanate.

The book introduces the reader to the history of the Islamic Caliphate and the Turkic peoples. This introduction is followed by annotated translated sources which illuminate the view of the Eurasian steppes in Muslim-Arabic geographical writing from the pre-Saljūq period, the self-image and ideology of the victorious Saljūqs and their fundamental claim to legitimacy, and the conventional narrative of the coming of the Saljūqs in later Arabic historiography.

Illustrating the variety of sources available on the history of Turkic tribes in the Eurasian steppes and in the central Islamic lands, ranging from geographical writing, to chronicles, to mythological legends, this book will be an essential resource for students and scholars with an interest in the Turks and their image, History, and Middle East Studies.

Yehoshua Frenkel is a Senior Lecturer in Middle Eastern History, University of Haifa, and is currently engaged in a research project entitled *The Environmental History of Mamlūk Bilād al-Shām*.

Routledge Studies in the History of Iran and Turkey

Edited by Carole Hillenbrand
University of Edinburgh

This series publishes important studies dealing with the history of Iran and Turkey in the period 1000–1700 AD. This period is significant because it heralds the advent of large numbers of nomadic Turks from Central Asia into the Islamic world. Their influence was felt particularly strongly in Iran and Turkey, territories which they permanently transformed.

The series presents translations of medieval Arabic and Persian texts which chronicle the history of the medieval Turks and Persians, and also publishes scholarly monographs which handle themes of medieval Turkish and Iranian history such as historiography, nomadisation and folk Islam.

Mediaeval Islamic Historiography and Political Legitimacy
Bal'ami's Tarikhnamah
A.C.S. Peacock

History of the Seljuq Turks
The Saljuq-nama of Zahir al-Din Nishpuri
Translated by Kenneth Allin Luther
Edited by Edmund Bosworth

The Annals of the Saljuq Turks
Selections from al-Kamil fi'l-Ta'rikh of Ibn al-Athir
D.S. Richards

Early Mongol Rule in Thirteenth Century Iran
A Persian Renaissance
George Lane

The Ghaznavid and Seljuk Turks
Poetry as a Source for Iranian History
G.E. Tetley

Iranian Elites and Turkish Rulers
A History of Isfahan in the Saljuq Period
David Durand-Guédy

Early Seljuq History
A New Interpretation
A.C.S. Peacock

The Great Seljuqs
A History
Aziz Basan

The History of the Seljuq State
A translation with commentary of the *Akhbar al-dawla al-saljuqiyya*
Clifford Edmund Bosworth

The Rum Seljuqs
Evolution of a Dynasty
Songül Mecit

The Turkic Peoples in Medieval Arabic Writings
Yehoshua Frenkel

The Turkic Peoples in Medieval Arabic Writings

Yehoshua Frenkel

Routledge
Taylor & Francis Group
LONDON AND NEW YORK

First published 2015
by Routledge
2 Park Square, Milton Park, Abingdon, Oxon OX14 4RN

and by Routledge
711 Third Avenue, New York, NY 10017

Routledge is an imprint of the Taylor & Francis Group, an informa business

British Library Cataloguing in Publication Data
A catalogue record for this book is available from the British Library

Library of Congress Cataloging in Publication Data
A catalog record for this book has been requested

ISBN: 978-0-415-74764-6 (hbk)
ISBN: 978-1-315-75295-2 (ebk)

Typeset in Times New Roman
by Saxon Graphics Ltd, Derby

MIX
Paper from responsible sources
FSC® C013056
www.fsc.org

Printed and bound in Great Britain by
TJ International Ltd, Padstow, Cornwall

Contents

Prologue

During the eleventh century, vast bands of Eurasian nomads, led by the Saljūq (in Arabic; Seljük in Turkic; Selçuk in modern Turkish) clan of the Turkic[1] Qïnïq[2] tribe crossed the Central Asian frontiers of the Abode of Islam and succeeded in wiping out all potential challengers. Ṭughril Beg, the son of Mikhā'il b. Saljūq, entered Baghdad and seized the reins of the 'Abbāsid Caliphate. A generation later, the Saljūq sulan Alp Arsalān led his followers westward across the Euphrates. They defeated the Byzantine armies and penetrated deep into Asia Minor, a land that in the future would be called Turkey. In the coming centuries, the Turks of Western Asia played a critical role in the challenges faced by various Muslim dynasties in this part of the world. They were conspicuous in battles against mighty enemies such as the Crusaders and the Mongols. A by-product of the Turks' significant military achievements was their ascendance to the throne in Cairo. They established a particular Turkic regime (*dawlat al-atrāk*) that governed Egypt and Syria. For a quarter of a millennium this manumitted slave regime, the Mamlūk Sultanate, played a major role in the history of the eastern Mediterranean. Subsequently, they were replaced by another Türk empire, that of the Ottomans.[3]

The coming of the "Barbarians of the Eurasian steppes" (*c.* 1030–1071)[4] is certainly a watershed in Islamic history, as a great number of historians agree. Surprisingly, only a handful of eyewitness accounts of these overwhelming events in eleventh-century Iran, Iraq and Syria have reached us. Nevertheless, later Arabic reports of fighting and diplomacy are a rich source for the study of how the Eurasians were viewed in the 'Abbāsid territories. These sources also shed light on the self-presentation of the Saljūqs. A selection of these texts forms this book's backbone.

This corpus is planned to serve as an introduction to the world of the Turkic people (*atrāk* in Arbic; *ṭurkāyē'* in Syriac) at a crucial moment in this people's history and to the Arabic narrative sources that shed light on their past.[5] The translated texts aim at providing a portrait of the complex relations between the pre-Islamic Eurasians of the steppes and the Abode of Islam, as well as of the changes in the way the *atrāk* were perceived by the population that inhabited the urban centres of the Middle and Later 'Abbāsid Caliphate (*c.* 833–1258).[6]

The structure of the book

Part I introduces the reader to the history of the Islamic Caliphate and the Turkic peoples. This brief introduction offers a basic acquaintance with the institutions, names, places and events that are mentioned in the various narratives. Part II is an anthology of translated primary Arabic sources that shed light on:

- the view of the Eurasian steppes in Muslim-Arabic geographical writing from the pre-Saljūq period;
- the self–image and ideology of the victorious Saljūqs and their fundamental claims to legitimacy;
- the conventional narrative of the coming of the Saljūqs in later Arabic historiography.

These accounts illuminate myths, images, social functions and organisation, political discourse, propaganda and historiography, which were formulated and articulated during the centuries preceding Ṭughril Beg's victories,[7] in the immediately following decades and during later generations.

The texts elucidate the triumphant Saljūqs' political claims, their self-representation, their ideology and their fundamental claims to legitimacy. Other texts embody the conventional narrative of the Saljūqs in Arabic historiography. One work provides a translation of a unique creation myth that prevailed among the Turkic peoples prior to their conversion to Islam. The selected texts also demonstrate the variety of sources that are at our disposal to study the history of the Turkic tribes in the Eurasian steppes and in the central Islamic lands. They run the gamut from geographical writing to chronicles and mythological legends.[8] The translations are supported by annotations as well as a bibliography for further reading at the end of the book.

Notes

1 Throughout this study the noun "Turkic" indicates the Eurasian nomads whom the Arab sources name *atrāk*.

2 *Kınık* in Modern Turkish. Maḥmūd b. al-Ḥusayn b. Muḥammad al-Kāshgharī (fl. 1072–1094), *Kitāb Dīwān lughāt al-turk*, facsimile volume in Besim Atalay (trans.), *Divanü Lûgat-it-Türk* (Ankara, 1941), 40 (l. 6): "The chief of them is Qiniq to which our present sultans belongs (*minhā al-salāṭīn fī zamānina*)" [Robert Dankoff (trans.), *Compendium of the Turkic Dialects* (Cambridge, MA, 1982–85), 1: 101]; Peter B. Golden, "'I Will Give the People unto Thee': The Činggisid Conquests and Their Aftermath in the Turkic World", *Journal of the Royal Asiatic Society* 10/1 (2000): 39.

3 For the pre-Islamic chapter in the history of the Inner Asian Turkic people, see Carter V. Findley, *The Turks in World History* (Oxford, 2005), 37–55, 66–9.

4 Clifford Edmund Bosworth, "Barbarian Incursion: The Coming of the Turks into the Islamic World", in D.S. Richards (ed.), *Islamic Civilisation 950–1150* (Oxford, 1973), 1–16.

5 For a later period, see Xavier Jacob (trans.), *Les Turcs au Moyen-Age* (Ankara, 1990); for the period under discussion here, see Şeşen Ramazan, *Islâm coğrafyacılarına göre*

Türkler ve Türk ülkeleri (Ankara, 1998), who selected a dozen Turkic (Kāshgharī), Persian and Arab authors.

6 For a different periodization and terminology, see Marshall G. Hodgson, *The Venture of Islam: Conscience and History in a World Civilization* vol. 2: *The Expansion of Islam in the Middle Periods* (Chicago, 1974), 3.

7 Compare to Claus Schönig, "On some unclear, doubtful and contradictory passages in Maḥmūd al-Kāšġarī's 'Dīwān Luġāt at-Turk'", *Türk Dilleri Araştırmaları* 14 (2004): 37–8.

8 Compare with Scott C. Levi and Ron Sela (eds.), *Islamic Central Asia: An Anthology of Historical Sources* (Bloomington, 2010).

Part I
Overview

1 The Turkic peoples (*atrāk*) and the Islamic Caliphate

The Turkic people first encountered the empires of the world of Late Antiquity when they created a new steppe empire (Türk Qaghanate) in Central Asia in the sixth century CE.[1] The territory that they inhabited was not unknown to Greek, Latin and Syriac authors. Historians of Byzantium added a new layer to the stories about the Eurasian peoples.[2] In the course of time, this topic was taken over by Arab authors who included new information on the Eurasian steppes and its population in their chronicles and geographical writings.[3]

Owing to the rapid expansion of Islam, Muslim troops soon reached the Caspian Sea and Inner Asia (*c.* 650). One of the results of these military achievements was direct contact between the Abode of Islam and the Eurasian peoples. The armies of the Caliphate engaged them in two frontier zones as early as the Umayyad period (660–750). One frontier (*ribāṭ, thaghr/thughūr*) was the land controlled by the Khazars.[4] The second frontier was in Central Asia,[5] in the territory that the Arab geographers called the "Land beyond the Oxus River" (Transoxiana = *mā warāʾ al-nahr*).[6] Turkic peoples constituted the lion's share of the population in this frontier land. Although the Arab geographer al-Iṣṭakhrī (d. 957) argues that the Khazars were not *atrāk*, nevertheless ceremonial rituals at the royal (*khāqān* or *bek*) court suggest the survival of Türk practices.[7]

2 Slaves and soldiers

Yet armed conflict was not the sole dimension of contact between the Eurasian peoples and the Islamic Caliphate. The political and military needs of the Caliphate created a hunger for qualified manpower to fill its armies and administration.[8] Demand for fresh conscripts kept the requirement for foreign recruits, including slave soldiers, high.[9] Among the recruits were soldiers from the frontier zones of the Abode of Islam, particularly from the Eurasian steppes. Within the rank and file of the new Islamic army two types of conscripts may be distinguished:

1 *Mawālī*: freedmen who were related or attached to an influential Muslim patron and were considered his clients.[10]

2 Slave soldiers (*ʿabīd*) who were enrolled by the Caliphate to serve in its new military battalions.[11] Some of these bonded men were *atrāk* (Turkic people).[12]

The assimilation of the Eurasian recruits into Islamic society took place in three stages: 1) enslavement and conversion to Islam; 2) promotion of slaves' descendants who became close companions of their masters or even retainers in the caliphs' service, filling high-ranking positions in the administration; and 3) integration into civil society.

The newcomers who arrived in the central Islamic lands developed into the nuclei of a new military force. In the new ʿAbbāsid army that regrouped after the wreckage of the war between al-Amīn and al-Maʾmūn, the sons of Hārūn al-Rashīd, the position of Eurasian page boys (*ghulām*, pl. *ghilmān*) became evident.[13] Their brother al-Muʿtaṣim pursued a policy of enlisting slave soldiers from the northeastern frontiers of the ʿAbbāsid Caliphate.[14]

Due to their growing numbers, social tension in the capital of Baghdad increased and led to violent incidents between the *atrāk* and the local population. This forced the caliph al-Muʿtaṣim to search for an alternative capital. He decided to move northward along the banks of the Tigris River and founded the city of Samarra.[15] The organisation of the city reflects the ethnic composition of the new army as well as the political role of the Eurasian soldiers. The geographical sources narrate that the caliph recruited Eurasian slaves and allocated them separate quarters, according to their lands of origin.[16] A growing number of them had a distinctive Turkic appearance and names.[17]

The story of the building of Samarra conveys some brief remarks that reflect the prevailing hostile attitude towards Eurasians, how they were perceived and their place in society. The historian and geographer Aḥmad b. Abū Yaʿqūb Isḥāq ibn Wāḍiḥ al-Yaʿqūbī (d. 897 or 905)[18] provides a long section on the foundation of this new ʿAbbāsid capital.[19] He narrates:

> In earlier times Samarra was a wasteland in the diocese of al-Ṭīrhān (Tīrhan).[20] Only a monastery stood in this region. There was no settlement. On the location of the monastery, the governmental (*sulṭān*) compound was built that is known as the general complex. The monastery building was converted to the treasury. In the year 218/833 al-Muʿtaṣim ascended the throne of the caliphate. He moved from the Mediterranean coastal town of Tarsus to Baghdad, and settled in the palace of his brother al-Maʾmūn. Then he constructed a palace on the eastern bank of the Tigris River. He dwelt there until the year 221/836. With him was a band of Turks. At that stage they still were barbarian infidels (*ʿajam*).
>
> Jaʿfar al-Khushkī told me: Al-Muʿtaṣim used to dispatch me, during the days of the caliph al-Maʾmūn, to Nūḥ b. Asad in Samarqand. He commissioned me with purchasing Turk slaves. Every year I returned to Baghdad, bearing with me a group of Turks. Already during the days of al-Maʾmūn he accumulated about three thousand pageboys (*ghilmān*). After he inherited the caliphate he strove to obtain more Turks. He brought slaves from private

> owners and thus collected a great number of slaves. Among them were Ashinas, who had been the slave (*mamlūk*) of Naʿīm b. Khāzim Abū Hārūn b. Naʿīm; Ītākh, who had been the slave of Sallām b. al-Abrash; Waṣīf, an armament smith (*zarrād*) who had been a slave of the al-Nuʾmān family: and Sīmā (silver) of Damascus, who had been a slave of Dhū al-Riʿasatayn al-Faḍl b. Sahl.
>
> Whenever these barbarian infidels rode they galloped on their horses and bumped into people without heed. Baghdad's mob often attacked them, killing and injuring some of them. It is in vain that they bled: No one took revenge and no perpetrator was punished. This troubled al-Muʿtaṣim and he decided to depart from Baghdad. First he travelled to al-Shamsiyya, a location at which al-Maʾmūn used to stay for long days and months. Al-Muʿtaṣim decided to build a town there, but the place was too close to Baghdad and the potential construction site was too small; hence he found the location unsuitable. His vizier al-Faḍl b. Marwān advised him in 221/836 to move to al-Bardān.
>
> One day, al-Muʿtaṣim rode and hunted until he reached Samarra, which at the time was a wasteland in the diocese of al-Tīrhan, bare from building or man. Only a monastery stood there. Al-Muʿtaṣim conversed with the monks. Inquiring about the place's name, he was told by a monk: "It is written in our ancient sacred books that this place is named Sāmarrāʾ, meaning: happy is the one who sees it. In olden days, Shem the son of Noah lived in this town. The place will be rebuilt after long years of desolation by a glorious, triumphant and victorious king. He will be accompanied by men whose faces resemble those of wild birds that live in dry deserts. He and his son will inhabit the town".
>
> Al-Muʿtaṣim cried out: "By God, I am the king. I will build it". He settled there, accompanied by his son. He narrated that one day Hārūn al-Rashīd instructed his sons to travel and hunt: "I went together with al-Amīn and al-Maʾmūn my brothers, and other sons of the caliph. Each one of us hunted an animal or a bird. I captured an owl. We returned and presented our prey. The eunuchs that accompanied us described to Hārūn al-Rashīd what each of us had caught. When it was time to explain what I had caught he saw the owl. The eunuchs were troubled by the possibility that the bird would escape or that I would be disgusted and evade presenting the owl. Hārūn al-Rashīd asked who had caught it and they said: 'Abū Isḥāq al-Muʿtaṣim'. Yet he regarded it as a good omen. He demonstrated his delight and rejoiced, saying: 'One day he will be the caliph. His army, his commanders and supporters will be people whose faces resemble the look of this owl. He will rebuild an ancient city and settle the place with them'".[21]

The recruitment of the *atrāk* did not save the ʿAbbāsid regime from disintegration. Moreover, *atrāk* officers and soldier became paramount players in the intrigues that tore apart the caliphs' courts.[22] Some of them grew to be kingmakers, while others were plotters and executioners. Hence, the role of the *atrāk* in the administrative apparatus of the Middle ʿAbbāsid Caliphate (833–934) cannot be ignored.

3 The Turkic tribes in the steppes

The formation of a long frontier between the Caliphate and the Eurasian steppes generated multidimensional relations between Islam and Central Asia. Among other things, this vast region exported a great variety of goods and materials to the Islamic centres in the Near East. Among the merchandise that reached Baghdad and other cities were Eurasian (Turkic nomadic pastoralist and other populations) men and women. Consequently, the ʿAbbāsid administration (*dīwān*) was keen to know more about the geography of these steppes. Clerks in the service of the Caliphate and scholars searched for information on the roads, provinces, cities, people and economy of this enormous area.[23] They compiled facts, figures and legends. Travellers added information.

Al-Masʿūdī (d. *c.* 956), for example, dedicated (in 943–944) a long chapter in his *Meadows of Gold and Mines of Gems* to the Eurasian peoples (the offspring of the biblical Japheth), including several Turkic groups. He mentions the disintegration of the Türk Qaghanate (*khāqān al-khawāqīn* in Arabic) and names several tribes: Kharlukh (Qarluq), Ṭughuz-Ghuz (Toquz-Oghuz, i.e., the nine tribal groupings), Kīmāk and Ghuzz.[24]

Already in the second half of the ninth century, Arab readers could consult a modest library about "the description of Earth" (*ṣurat al-arḍ*). The texts included in this genre are not merely a report on lands (*bilād*) and nations (*ajnās*). Using oral and written reports together with imaginative topography, these literary writings were instrumental in explaining the *mappa mundi*. Such, for example, are the narratives of Islamisation, both of people and places. Another theme in these writings is the description of the barbarians' abode in Eurasia.[25]

From these sources and from Arabic chronicles we can deduce that during the closing decades of the eighth century, Turkic tribes and confederations, affected by the collapse of the second Kök Türk Qaghanate (Gök Kağanlığı in modern Turkish; fl. *c.* 545–744), moved westward towards the frontiers of the Abode of Islam. Among these Eurasian steppe tribes were the Oghuz tribes (see page 000).[26]

4 The image of the Turks in ʿAbbāsid-period writings

Various Arabic literary genres transmit data on the Turkic peoples and their lands (the Eurasian steppe). Some of the records viewed the remote territories that stretched beyond the frontiers of Islam as the fringe of mankind. The writers regarded the people inhabiting them as bizarre creatures and the lands that they inhabited as fantastic domains. From these traditions we may deduce that at this time the prevailing attitude towards the Eurasians among Muslims was unfavourable.[27] A clear depiction of the *atrāk* of the steppes and their image is provided by a letter allegedly sent by Aristotle to Alexander the Great.[28] In it, the famous philosopher is said to advise the victorious king:[29]

> O Alexander! Be careful not to hasten and attack the Turks. Dispatch the Sogdian troops against them because they are familiar with the Turks and

> know better than the Greeks how to fight them. Acquire the advice of the people of Khurāsān [Khorasan]; you will benefit from their stratagem because the Turks are not familiar with their devices. They are not familiar with night warfare. So use this tactic against the Turks. Advance towards them quietly with no noise. Do not act unjustly and do not act arrogantly, because injustice renders determination useless. If the war between you and them is prolonged, enslave a group of them. Be careful, treason is most common among them. [...] Among the Turks there is no infantry, their lives depend totally on their beasts. [...] They enjoy food, drinking and amusement and are satisfied with it; they do not like money and do not desire riches. [...] Watch out, the Turks are not scared of death nor fear being killed, but they worry about imprisonment. If you take them captive, be lenient and manumit them. Provide them with the best clothing, because this is very important to them. When you manumit them, be generous and provide them with food and drinks. They like it very much and desire it. They consider these gifts as a magnificent fortune.[30]

Some sayings attributed to the Prophet Muḥammad (*ḥadīth*) reflect the harsh image of the *atrāk* among the educated urbanised Muslims in the early and middle ʿAbbāsid periods. These traditions perceived the *atrāk* as deadly dangerous enemies of Islam. The suspicious attitude towards the *atrāk* is drawn in apocalyptic terms. A well-known *ḥadīth* says: "The Prophet said the *atrāk* will evacuate the people of Iraq from their residences [and supersede them]." Another tradition argues in the name of the Prophet: "The hour of the resurrection (*sāʿa*) shall not come until the Muslims win a decisive victory over the *atrāk*. These are people with faces resembling iron shields, who wear furs. The Prophet pointed to the East and said: From this direction the horn of Satan will emerge."[31] Muslims were advised: "Leave the *atrāk* (*tarik[u] al-turk[a]*) as long as they do not trouble you."

Similar sayings are ascribed to Companions of the Prophet. The caliph ʿUmar b. al-Khaṭṭāb is said to have written to his governors: "The *atrāk*s will conquer Kūfah. If you meet a Turk kill him. Do so because after the year 200/826 they will revolt." In his *Akhbār al-ḥamqá wal-mughaffilīn*, Ibn al-Jawzī (1117–1201) narrates the following episode: A person heard an *imām* recite from the Qurʾān. At chapter 30, *sūrat al-Rūm*, instead of the canonical text "the Byzantines have been defeated" (verse 2), the reciter read: "Did not the *atrāk* win?" So Ibn al-Jawzī asked the reader why he changed the lines. The answer was simple and blunt: "They all are the enemies and it is not important who amongst them is mentioned."[32]

Fighting against the Turks is depicted as a very risky enterprise. Only a person who is ready to sacrifice his life for the cause of Islam would involve himself in this act that requires bravery. This vision of the risks immanent in facing the Turks is clearly reflected in an account of Ḥusayn b. ʿAlī. According to this, when the courageous grandchild of the Prophet Muḥammad was besieged by the Umayyad army at Karbalāʾ, he suggested to his adversary: "Send me to the front to fight the *atrāk*, I shall fight them till my death."[33]

This popular image of the *atrāk* is echoed in anecdotes that were integrated in historical writings. In some of the accounts the *atrāk* are depicted as violent creatures and as a source of danger to the settled population of the Fertile Crescent.[34] The negative representation of the *atrāk* as new barbarians is visible in reports about the Eurasians that reached to Baghdad and other centres. In an account of a deliberation regarding the characters of various people (*umam*) the Turks are accused of being wild beasts of prey (*sibāʿ al-ḥirash*).[35]

5 Gog and Magog

The negative image of the *atrāk* fitted the discourse of medieval Arab and Persian writers, who employed several instruments to delineate mankind. One of them was the use of biblical and mythological writings as a source to explain the ethnic and cultural diversity of mankind. In this scheme of humanity, various mysterious peoples represented the Turkic peoples.

Pre-Islamic tradition already identified the Turks with the biblical Gog and Magog, who also in Islamic tradition are depicted as a severe threat to the welfare of the rest of mankind.[36] This cultural construction of mankind is based on a popular interpretation of Ezekiel 38,[37] and it became common also in medieval Arabic writings.[38] Both Muslim and non-Muslim authors claim that the imaginary Gog and Magog were confined by Alexander (Iskandar)[39] behind a barrier (*radm*),[40] a story that can be traced in the Talmud and also in exegeses of the Qurʾān.[41] It is said that the Hellenistic emperor built this wall to prevent these hidden forces from attacking the civilised world.[42]

Some Muslim scholars located the *radm* at the northern edges of the inhabited world. They did not agree upon the exact borderline. Some located it at the end of the steppes, others placed it on the frontier of Khazariyya, namely in the Trans-Caucasian plains. Both lines of interpretation agreed that the *atrāk* were so named because they are "those who were left [*turiku*] behind the barrier".

The construction of this barrier is a major topic in the medieval Arabic legend of Alexander. Ibn Khurrdādhbih transmits one of its earliest versions:

> On the right side of the Tughuz-Ghuzz, towards the south, is Tibet. After defeating Fur (Porus), the monarch of India,[43] Alexander (Dhū al-Qarnayn)[44] executed him. He remained in India for seven months, dispatching armies against Tibet and China. One of the commanders that he had dispatched returned to his court and informed him that the kings of the east wished to be Alexander's subjects. Learning that the two kings of Persia and India, Dārā (Darab the Younger or Darius)[45] and Fūr, had been defeated by him, and knowing his justice and noble conduct, they would pay him tribute.
>
> Alexander (Iskandar) departed India where he had installed a governor he trusted and who he provided with thirty thousand men. He traveled to Tibet and was met by the king who came accompanied by his nobles (*ṭarkhan; ṭarxan*)[46] to convey his surrender. The king stated: "O magnificent Alexander, I learned of your justice and your fidelity, even towards those who rose up

against you. Although I do not know if all your victories result from the Almighty's wish,[47] I do not wish to obstruct you in accomplishing your aspiration, nor to fight against you. Whoever challenges you, fights against you or defeats you would achieve it only because God's hand holds victory. Whoever struggles against God will be defeated. I, my people and my authority are at your service. Command us and we shall execute your orders".

Alexander responded favourably and said: "Those who know God's truth must fulfill His commandments. We are obliged to execute His precepts. I hope that you will discover the genuineness of our justice and will be satisfied with our loyalty". Alexander also asked for the king's advice with regard to the Turks of the steppe, because the sedentary Turks had already submitted to his rule. He travelled ahead of him. The king of Tibet offered Alexander numerous gifts, but the latter declined them. Yet the king did not stop urging Alexander till he gave in and agreed to accept the presents. The king brought to him four thousand donkeys burdened with gold and a similar volume of musk. Alexander handed a tenth of the musk to his wife Rushānak (Roxane, Ruqayya), the daughter of Dārā the king of Persia. The rest of it he distributed between his leading officials. He deposited the gold in the treasury.

The king of Tibet asked Alexander to nominate him as the commander of the vanguard that would head to China. The proposal was accepted by Alexander, who ordered the king to appoint his son as deputy and he nominated his son Mudābīk (or Mudābīl) as viceroy of Tibet. The king of Tibet was joined by one of Alexander's top officers, who commanded an army of ten thousand soldiers. They advanced to China, followed by Alexander with the bulk of the army.

At the head of ten armies, each of them comprising one hundred thousand soldiers, the ruler of China advanced towards the invading armies. He had sent an embassy who admitted that Alexander enjoyed a very positive image, that he was renowned for his loyalty and honourable deeds. The embassy added that because of this their king refrained from fighting him, although if he opted to fight he would not fail. Hence he asked Alexander to elaborate his demands, so that he would be able to fulfill them. Alexander accepted this proposal and demanded that the king of China, similarly to his custom in other lands, pay him a yearly tribute of a tithe. He threatened to fight the king if he failed to do so, saying that the size of the Chinese army did not terrorise him. "Allah will help the small party to defeat a considerable army".[48]

Alexander sent this response with a mixed delegation of Persians and Indians. He advised them to tell the king about his generosity and his justice, as well as about his attitude towards the inhabitants of the countries within his realm. The king of China was persuaded to recognise Alexander's lordship and he suggested paying the tribute and buying peace with *firind* (silk, or figured cloth),[49] swords and other tools. Alexander was pleased with the response and agreed to the king's suggestion. He demanded one million *firind*, one million pieces of silk, half a million (pieces) of brocaded silk, ten thousand complete sets of riding equipment: saddles, bridles, belts, stirrups and other components, and one million maunds of silver. The king of China paid it. Alexander remained in the territory of the king of China till he completed the construction of a town named Burj al-Ḥajāra (the Stone Citadel). He manned the fortification with five thousand Persian fighters and commissioned one of his generals, named Nukilidis, to head these border guardians.

From there Alexander continued his journey. Accompanied by the king of China, he entered the country of Shūl (Ṣul/Chul) and conquered his territory. He built two towns there, one named Shūl and the other (called) Khamdān. Alexander ordered the king of China to encamp with his army in Khamdān and to station a battalion of frontier guardians in Shūl. He himself advanced towards the Turkic groups of the steppe.[50] Alexander proceeded in his campaign till he conquered their land and subdued the Turks. There he learned about an enormous number of Turkic peoples who lived in the northeastern corner of that continent. They were said to be wrongdoers. Alexander consulted the king of China who explained that these Turks do not possess valuable property that can be plundered. They only have herds and iron. From the north the Green Ocean closes on them. No one can cross this water. From the west and the south, high inaccessible mountains close on them. No one can travel across these terrains that reach up to the sky. Those Turks can use only a single path, narrow as a strap. They are locked in a corner and if this road were to be blocked, they would be caught there. This will relieve mankind and will stop their harassment of humankind.

Alexander perceived the soundness of the king of China's wise advice and built a barrier that blocked this ravine in the mountain. This is the barrier that the Almighty mentions in the Qur'ān and describes.[51] Then he, the Two Horned Alexander, returned to the land of the Turks who dwell in towns and who are polytheists and crossed it till he reached the land of Sughd (Sogdiana).[52] There he built the towns of Samarqand, al-Dabusiya and Alexandria (al-Iksandarīya) the Remote. Alexander advanced further to Bukhara and built there a town named Bukhara. In the country of Marv he built a town that is named Marv, as well as the towns of Herat and Zarnaj. He left Transoxiana via Jurjan. In Iran he commissioned the construction of the towns of Rayy, Isfahan and Hamadhān. Eventually he arrived in Mesopotamia, where he lived for several years.[53]

Several Arabic geographical writings from the middle ʿAbbāsid age have transmitted a story concerning the caliph al-Wāthiq (r. 842–847), who is said to have sent his chief interpreter, Sallām al-Turjumān (the dragoman), to inspect a barrier, which presumably is located in the Caucasus. Returning to the caliph's headquarters in Samarra, Sallām wrote an account of his mission to the Trans-Caucasian steppes, which was partly preserved.[54] His narrative was copied by later writers and incorporated in works that reflect the place of the Turks in the popular Islamic *mappa mundi*.[55]

6 Literary ethnicity (*shu'ūbiyya*)

The constant flow of Eurasian people to the central Islamic lands generated among its population a multifaceted reaction towards the Turkic peoples. It seems that zealous loyalty and partisanship (*'aṣabiyya*) played a major role in public discourse throughout the lands of the Caliphate. Each ethnic type was classified and its hypothetical role was set.[56] Some authors added remarks on the physiognomy and the way of life of the Turkic peoples to their geographical descriptions.[57]

The ethnic conceptions that prevailed in the lands of the 'Abbāsid Caliphate are clearly reflected in the *shu'ūbiyya* controversy.[58] When writing on people and lands, 'Abbāsid authors often used the literary device of a presumed debate. This tool enabled them to present conflicting reports on their subject-matter. The best-known version of the pseudo-debate literature about the qualities of the Turks is the piece by al-Jāḥiẓ (d. 869). In his epistle on the *atrāk* soldiers, al-Jāḥiẓ extolled their high merits.[59]

Al-Jāḥiẓ was not a unique voice among the 'Abbāsid writers who dwelled upon the virtues of the Turks. In the generations that followed him, several authors argued that the *atrāk* are excellent riders and archers, and they were portrayed as courageous and loyal people.[60] *Atrāk* men and women were renowned for their beauty, stamina and valour.[61] The Turkish slave girl was the stereotype of the ideal slave.[62] Furthermore, the pseudo-ethnological discourse did not lose steam many years after the heyday of the *shu'ūbiyya* controversy, as a quick review of chronicles and inscriptions from the Mamlūk and Ottoman periods would reveal.[63]

Anecdotal narratives convey historical information and social values.[64] They inform about cultural images. Clear examples of ethnic stereotypes that prevailed among the ruling elite during the middle 'Abbāsid period can be easily traced in al-Tanūkhī's writings. The well-known littérateur represents the *atrāk* officers as generous simpletons. He describes a pseudo-occurrence:

> Abū al-Ṭayyib Muḥammad b. Aḥmad b. 'Abd al-Mu'min, the officer at the courts' gates in the city of Ahwāz, told me that a Baghdadi debtor had narrated to him about one of their sheikhs who became extremely rich till he gave up begging. This respectable person taught them how they should perform.
>
> He said: "We asked him the reasons for his generosity and he told us the following story: 'I learned Syriac till I gained the capacity of reading the Christians' prayer books. Then I dressed a hermit's gown and went to the city of Samarra in northern Iraq, where the *atrāk* army commanders were dwelling. I approached a mansion and asked its resident's permission to enter. He invited me in and I told him that I came all the way from a distant monastery in Syria, where I had served for over thirty years as a monk. One night I saw in my dream the prophet Muḥammad entering my chapel and calling me to convert to Islam. I responded to the Prophet's call and he instructed me: "Go to this officer, who is among the persons that will arrive in Paradise, so that he will be present during your conversion". This is the reason for my arrival

at your house. The Turk officer was very pleased to hear my account; however he stammered and was not capable of receiving my conversion. So I cut my girdle and accepted Islam in front of him. He bestowed upon me gifts, clothes and money in the total value of five thousand silver dirham coins, and I returned to my lodgings.

'Early next morning, dressed as a monk, I went to the residence of another Turk commander and told him the same story that I had told in the previous day to the first officer. He gave me more gifts. So I wandered from one house to another telling my story and gaining gifts. I collected a considerable sum of more than 50 thousand silver coins.

'Several days passed and I went to the house of one of the *atrāk* army commanders. It happened that a gathering was taking place there. Several of the leading *atrāk* participated in it. I told them my story and succeeded in pleasing them. However, amongst the audience was an officer whom I had already met and hence was familiar with the tale. Following my presentation and conversion to Islam, the Turk officer bid me farewell. Yet, the wheel of fortune swung against me.

'As I left the gathering the Turk officer, whom in the past I already had met, followed me, accompanied by his slave. He caught me outside the house's gates and carried me to the mansion of the Turkish officer I had visited first. I was in turmoil. Sensing that it was my worst moment I offered the slave a substantial fortune, promising him that if he freed me I would give him all the money and goods that I had collected. But the slave declined my offer.

'The Turk officer came, evidently he was devastated, and approached me: "O daddy: you were mocking the *atrāk*, making every single amongst them a joke and collecting their money". Trembling in fear I said: "O sir, I am a poor man, a plagiarist, a swindler, I cheated in order to gain something". The Turk responded: "Are you afraid that I will make you a subject of mockery and ridicule you in your hometown? I am not this sort of man. You were successful in deceiving me as you were doing well in misleading the entire officer corps. However, who are you?"'

The swindler continued his story: "I calmed him down and made him happy with slapstick and jesting that amused him and caused him to laugh. He ordered to bring wine and drank while I was entertaining him. My spirit improved. I was jailed in his house. After a while he provided me with gifts and money. He invited a group of Turkish officers to his house and I performed in front of them, dressed in swindlers' costume. Noisily they joked with me and had fun. The Turk officer told them my story and they were amused".

"In these circumstances I collected from them a considerable sum of money and returned to Baghdad, where I bought real estate. From this property I make my living".[65]

In the tale translated above, the *atrāk* play the role of high-ranking army officers who are tricked by a professional swindler. This image combines two opposite characters: that of devoted Muslims who enthusiastically support the spread of

Islam and that of gullible commanders. Yet the anecdote also indicates the process of acculturation that these Turkic people experienced and the deep change in representation as dangerous savages.

7 The Eurasian steppes in the ʿAbbāsid Geographical Library

One of the earliest chapters on Eurasia and the Turkic peoples was composed by ʿUbayd Allāh b. ʿAbd Allāh Ibn Khurradādhbih [or Ibn Khurdādhbih] (*c.* 820–916), who served in the communication and internal intelligence bureau of the ʿAbbāsids. In this capacity he wrote, quite early (starting in 844), a condensed account of the provinces, highways, cities and taxation regime of the Caliphate (*al-Masālik wal-mamālik*). Although the book contributes next to nothing on the human geography of lands beyond the borders of the Abode of Islam, including Central Asia, nevertheless it illuminates the conditions on the steppes' frontiers and the early image of the Eurasian people (the so-called *atrāk*), whom the Muslims envisaged as a fearsome threat.

Moreover, Ibn Khurradādhbih's work is often quoted by Arab writers, who transmitted especially those unique accounts that cannot be found in other sources. An example of this is his transmission of a report by Sallām the dragoman, whose expedition into Inner Asia was mentioned earlier.

Ibn al-Faqīh al-Hamadhānī

Another early account of Eurasia and the Turkic peoples was composed by Aḥmad b. Muḥammad Ibn al-Faqīh al-Hamadhānī. Information about him is meagre.[66] From the adjective al-Hamadhānī, it is possible to deduce that he was born and raised in western Iran, in the city of Hamadhan. Based on internal and external evidence in his *Kitāb al-Buldān* (The Book of Lands and Countries),[67] it is likely that Ibn al-Faqīh produced his work no later than 903, maybe during the first years of the caliph al-Muktafī (902–908).

Although criticised for his style,[68] Ibn al-Faqīh should not be rejected by students of Eurasian history and the Turkic peoples. Despite its shortcomings, his work provides the reader with rich information on medieval Central Asia. His contribution rests upon his role as transmitter of earlier accounts from the Central Asian frontiers that have not survived independently.

Ibn al-Faqīh opens the chapter on the *atrāk* by transmitting several prophetic traditions that reflect the common negative stereotypes of the Turks among the urbanised Muslims. These *ḥadīth*s reflect a deep sense of anxiety about the Turkic peoples. Next, Ibn al-Faqīh moves to the frontier between the Caliphate and Eurasia, informing his readers about the land, its population and trade.

He incorporates a report by Tamīm b. Baḥr, who visited the Uyghur capital on the Orkhon (Mongolia). This journey took place at an uncertain date (after the death of Hārūn al-Rashīd in 809; most probably towards the end of the first quarter of the ninth century). Since Tamīm mentions the horses that the ruler of the Turks provided him, it seems that he was on an official visit. The report includes eyewitness accounts and quotations from informants. Also, it briefly narrates the topography and geography of Inner Asia, ethnography, politics and fighting.[69]

Other Arab travellers followed in the footsteps of Tamīm b. Baḥr and penetrated deep into the Eurasian steppes. Ibn Faḍlān departed Baghdad in the days of the caliph al-Muqtadir (in 921).[70] Abū Dulaf Misʿar b. al-Yanbūʿī al-Khazrajī (fl. 943–952) wrote two travel accounts, one of which reports his travels across Eurasia.[71] The short text is often vague, and more likely is a fabrication of a narrative aimed to please the reading public. It is worth mentioning that the interest of Arab travellers in the Eurasian steppes did not stop after the establishment of the Saljūq Sultanate.[72]

The geographical accounts from the high days of the Caliphate visibly reflect the prevailing image of Central Asian nomads in the minds of the ʿAbbāsid elite. The Turkic peoples were portrayed as a menace that threatened the very existence of mankind, as well as superior warriors in the cause of Islam. This pseudo-anthropological approach did not lose steam many years after the high days of the *shuʿūbiyya* controversy between Arab and non-Arabs. The language of contempt that sedentary civilisations often demonstrate towards nomads would explain the tone of writers such as Bayhaqī, al-ʿUtbī and other eleventh-century Muslim authors.

8 The Iranian Intermezzo

Since a great part of Ibn Ḥaṣṣūl's epistle is concerned with the revocation of al-Ṣābī and his history of the Buwayhids (or Buyids), it is worth making a short detour to discuss briefly the history of this dynasty and their role in tenth-century Iran and Iraq.

Due to natural obstacles, the armies of Islam encountered firm opposition during their efforts to penetrate the Caspian provinces of Mazandarān (Ṭabaristān)[73] and Jīlān (Gīlān), and the neighbouring mountainous region of Daylam.[74] The first deep penetration of ʿAbbāsid forces into Ṭabaristān occurred only during al-Manṣūr's caliphate (r. 754–775). Effective control was achieved only during the reign of al-Muʿtaṣim (r. 833–842).

The harsh conditions in the mountainous region of Daylam produced a tough people.[75] They spoke their own dialect and clung to their traditions and ancient political institutions. This led Muslim authors to consider them hill-billies and uncouth people.[76] A shortage of economic prospects encouraged young men to move from these remote highlands to the Iranian plateau. There they often were recruited to serve in the armies of governors.

The change in the fortune of the people of Daylam occurred in the tenth century. At this historical juncture, three brothers, ʿAlī, Aḥmad and Ḥassan, the sons of Buwayh (hence Buwayhids), appeared on the stage of central Iran.[77] After a short while, they headed south to the rich province of Fars (in 322/934). The disarray in Iraq only improved their odds. Aḥmad b. Buwayh advanced to Ahwāz in central Iran (941). A few months later he entered Baghdad without encountering any opposition. The caliph conferred the royal rank of Muʿizz al-Dawla ("the Powerful who strengthens the [ʿAbbāsid] Dynasty") on him.

The Buwayhid regime was based on a family coalition and a semi-feudal administration (*iqṭāʿ*).[78] To fortify their political legitimacy, they advanced two conflicting accounts of their origin. On the one hand, they claimed to be the offspring of Bahram Jūr (Vahram Gur son of Yazdegird).[79] On the other hand, they also pretended to belong to the Arab tribe of Banū Ḍabba and boasted of themselves as being superior to other nations. They also maintained that they converted to Islam voluntarily.

In an effort to gain public recognition as legitimate rulers, they adopted ancient Iranian titles.[80] Simultaneously, they used Islamic titles that signified their support of the ʿAbbāsid dynasty (*dawla*). The Islamic political thought of that age prevented them from grasping the reins of government directly. The Sunnī jurists prescribed that the leader of the nation of Islam (*umma*) should be a true Arab from the lineage (*min al-ṣamīm*) of Quraysh.[81] This might be one of the reasons that stopped the Buwayhids from dethroning the ʿAbbāsid *amīr al-muʾminīn*. However, although they ruled in the name of the ʿAbbāsids, they were quite independent.

9 ʿAḍud al-Dawla

Abū Shujāʿ Fanna-Khusrau (born in 936), who is better known by his honorific title (*laqab*) ʿAḍud al-Dawla ("the Vigorous Arm who upholds the [ʿAbbāsid] Dynasty"), was probably the most skilled and capable Buwayhid ruler (*amīr*; *padi-i shāh*). The income from the rich province of Fars helped him muster a sizeable army.[82] He erected many buildings in his capital, Shiraz.[83] The inauguration festival of his palaces was predominantly associated with the pre-Islamic glory of Sasanian Iran. An inscription engraved on the walls of Persepolis (Takht-e Jamshīd)[84] presents his goals, as well as his interest in the pre-Islamic past of Persia.[85] This is echoed in the poems written by the famous al-Mutanabbī (905–965), who in several of his eulogies sang the glory of his master, the King of Kings (*shāh-an shāh*).[86]

Aiming to secure his lofty position, the *amīr* Abū Shujāʿ Fanna-Khusrau demanded that the caliph al-Muṭīʿ (r. 946–974) bestow upon him the title Tāj al-Dawla ("the Crown of Sovereignty"). However, the claim was rejected at this stage, yet after a while (in 962) the caliph gave in and granted him the desirable royal title ʿAḍud al-Dawla. The letter of appointment was accompanied by gifts.[87]

Showing remarkable military talent, ʿAḍud al-Dawla seized the province of Kirmān. Next he landed on the beaches of Oman (ʿUmān) on the eastern coast of Arabia (974).[88] The geostrategic location of Oman granted him partial control of the route from the Persian Gulf to India. His military prowess and financial resources gave the young Buwayhid prince the necessary confidence and backing to claim his family's prized possession, namely the post of *amīr al-umarāʾ* in Baghdad (January 975). Yet, despite his remarkable achievements ʿAḍud al-Dawla was forced to relinquish his position (June 975). However, this retreat did not end his aspirations.

The second round of skirmishing took place after the death of Rukn al-Dawla, the eldest Buwayhid, in September 976. ʿAḍud al-Dawla returned to dwell in the royal residence, next to the caliph's palace (978). He obtained additional royal titles: as well as *amīr al-umarāʾ* (the commander in chief; 981),[89] that of *tāj al-milla* (Crown of the Muslim Community) and *shāh-an-shāh* (King of Kings).[90] Both titles reflect the ancient Iranian tradition of the great kings. His political authority and economic power were unlimited.[91] He was considered by some to be the greatest king (*al-malik al-sayyid*) of his time.

With Baghdad under his jurisdiction, ʿAḍud al-Dawla gave his mind to reorganising the state bureaucracy and taxation. He was then free to turn his attention to the Jazira (upper Mesopotamia; the "island" between the Euphrates and the Tigris). A triumphant military campaign brought him to the borders of northern Syria. He even took to the field and led a holy war (*jihād*) campaign against Byzantium.

ʿAḍud al-Dawla claimed to fulfil the role of guardian of the ʿAbbāsid dynasty. Presenting himself as the only one who could safeguard the survival of the Sunnī Caliphate, he offered his daughter to the caliph al-Ṭāʾiʿ in 979.[92] Although ʿAḍud al-Dawla manifested his devotion to the welfare of the ʿAbbāsid dynasty and the Sunnī world vision,[93] and made efforts to contain Fāṭimid-Ismaʿīlī activity,[94] he actually reduced the caliph's title to a mere symbol.

ʿAḍud al-Dawla spent money and effort in fashioning a court that would radiate glamour and power.[95] He invested vast sums in building construction. His privileged position is reflected by the fact that he and the caliph's vizier (*wazīr*) agreed that no letter should be sent from the royal chancellery unless both notables confirmed it and the vizier appended his signature. In these circumstances it is not surprising that he asked Abū Isḥāq Ibrāhīm b. Hilāl al-Ṣābī (d. 994) to compose a book on the reign of the Buwayhid dynasty and the people of Daylam.[96]

10 The coming of the Eurasian barbarians

With the decline of the Buwayhids' authority, the weakened ʿAbbāsid Caliphate was ripe for another cycle of occupation. On this occasion the new forces were Turkic nomads led by the Saljūqs (a household from the Qïnïq Oghuz tribe), who had only recently gained access to the land of Islam.[97]

Arab and Persian authors mention several Turkic groups that roamed the Eurasian steppes.[98] In the closing decades of the tenth and early eleventh centuries, it is possible to identify three types of Turkic forces in the Central Asian frontiers of Islam: 1) nomads who operated on both sides of the frontiers; 2) tribes that were recruited by Islamic local political entities; and 3) *atrāk* invaders who seized power in countries within the Abode of Islam.

While imagining nomads, we should bear in mind that these peoples wandered constantly. Indeed, from the fragmented information on the population of Eurasia, it is clear that the ethnic and demographic maps of this vast continent were not static, rather the contrary. Displaced by climate fluctuations and wars, these peoples were on the move frequently. One direction of their advance was towards the sedentary regions south of the steppes, where the attractive domain of the Abode of Islam lay poorly protected.

The Oghuz tribe (Ghuzz or *al-atrāk al-ghuzziyya* in Arabic sources) was one of these Eurasian peoples. Presumably, their name defined a political coalition composed from several Turkic tribes.[99] The territory of the Oghuz/Ghuzz stretched, according to ʿAbbāsid-period geographical writings, from Lake Bakhash to the Syr Darya (Jaxartes). It seems that the axis of their wandering was in the desert (Kizil Kum = red sand) east of the Caspian Sea, and the steppes adjacent to the lower Volga. "The desert is their father and mother, as the cities are for us," wrote a Ghaznavid military commander, "they [the *turkmānān Saljūqiyān*] are scattered in the open land in small groups of twenty or thirty men".[100] Like other steppe nomads, the Ghuzz too were well-trained troops, and fighting warlords recruited them. This brought them across the Oxus River to Khurāsān.

In the first half of the eleventh century, the relations between the Inner Asian steppes and northern Iran changed drastically. The collapse of the Samanids [101] opened the door to Turkic tribes that gradually penetrated the Abode of Islam. The sedentary civilisation of Islamic Central Asia came under heavy pressure from these nomadic forces. This happened at the time when three Turkic forces were successful in reshaping the political order on the eastern fringe of the Islamic world.

A succession of victories made the Ghaznavids key players in Iran, Afghanistan and India (975–1187). Yamīn al-Dawla and his son Masʿūd held supreme power in the eastern Islamic world at the time the Saljūqs penetrated the arena. Manumitted Turk slaves were the nucleus of this dynasty.[102]

The nucleus of the Qarakhanid state (the Ilek or Ilig-khān Qaghanate, 999–1127) was the Qarluq Turk tribal confederation. Islam had already gained followers among them at the beginning of the tenth century, when Sabuq (or Satuq) Bugha Khān (d. 955) [103] converted to Islam and even assumed the Arabo-Islamic name of ʿAbd al-Karīm.[104] Led by Shihāb al-Dawla Hārūn b. Sulaymān b. Bugha Khān (d. 993), who bore the title of King (Ilek) of the Turks and Lord of Kashghar, the Qarakhanids gained control over Ferganah and Bukhara. They acknowledged the authority of Baghdad inasmuch as the ʿAbbāsid caliph was the only source of legitimacy. Their armies were composed of a small troop of retinues (*ghulām*, pl. *ghilmān*) and a large body of Turkic tribesmen (*turkmān*).[105]

Towards the closing of the tenth century, a new player broke into this arena. Growing numbers of Turkic nomads moved from the steppes (Turkistan) and erected their tents deep in the soil of the Abode of Islam. Among the different Turkic tribes who reached the boundaries of Transoxiana and Khwarizm were the Ghuzz. They would be the principal players in the next decades.

It seems that a combination of two conflicting powers explains the arrival of the Ghuzz and their military success. On the one hand, as early as 986, the nomads were attracted by potential grazing for their herds. This was a rough time. Northern Iran was afflicted with a grave famine that lasted several years. Nishapur was stricken by heavy snowstorms in the winter of the year 1010. The cold winter was followed by calamity and heavy famine that spread all over Iran and Iraq

(1010–1011). A great number of people perished. On the other hand, the political instability in Islamic Central Asia created a demand among the local rulers for trained cavalry and soldiers.

11 The Saljūqs

It is not clear who, at this historical moment, headed the Turkic tribesmen within the provinces of Sughd (Sogdiana). The vague early accounts shed no clear light on the difficulties we face in our efforts to reconstruct the history of these years. One source names the obscure al-Ḥasan (or al-Muḥsin) ibn Ṭaq as the head (*raʾīs*) of the advancing Turks. However, after a while the sources speak about bands of Turkmans led by the house of Saljūq (Seljuk) from the Qïnïq tribe.

Hence, before we continue with the chronology of the events, a few words on historiography. The image of the Saljūqs in the records composed by courtiers of the rulers of Iran and Central Asia is negative. They were depicted as vicious barbarians who had betrayed the legitimate rulers of the eastern Islamic world. The startling military achievements of the Turks in the Iranian plateau intensified the view of the steppes as a source of danger to the civilisations of Islam.[106]

The Saljūqid sources, however, paint a completely different picture. Following the conquest of Baghdad in 1055, a new political order emerged in Western Asia. The heads of the Turkic tribesmen seized de facto authority in this part of the Abode of Islam. Authors operating in the bureaucratic apparatus of this new Islamic power produced their writings after their patrons acquired the lofty designation of the Sultan of Islam. Subsequently, they adopted a historical narrative that improved the image of their masters.

Authors in the service of the early Saljūq sultans argued that their masters' forefathers played a significant role in the steppes already prior to their conversion to Islam.[107] These sources use the past as a component in the legitimisation of the new regime.[108] The Saljūq household is said to be amongst the retinue of the Turkic king (*oruq*). From other sources we learn that Duqāq or Tuqāq,[109] the father of the dynasty, was in the camp of the king of the Khazars. It was claimed that the title *beg* (tribal chief), attached to the personal names of male members of the Saljūq family, reflects the aristocratic position of the Saljūqs in the steppes.[110] Arsalān-Isrāʾīl's title of *yabghu* is said to be another indication that the Saljūq family belonged to the Turkic aristocracy.

The widespread version of the events says that Isrāʾīl-Arsalān took his men to Transoxania. Other Ghuzz clans, some of them headed by Mikhāʾil b. Saljūq, wandered between Bukhara and the Kara Kum desert (*c.* 1025). After a short time they crossed the Oxus River and migrated to Khurāsān (1027), only to be defeated by the Ghaznavids in 1029. Peace was a short-lived enterprise, as Shihāb al-Dawla Masʿūd (r. 1031–1040) is reported to have taken violent measures against Turkic tribesmen.

Turkic tribes (*turkmānān wa-saljūqyān*) that roamed in the steppes north of the Amu Darya (Oxus) River reached Nasa (Nisa; contemporary Ashkhabad). They are said to have numbered ten thousand cavalrymen led by Pighu (Bayghu?)

Mūsā. In addition to him, two other Saljūq commanders are named: Ṭughril (Toghril, Tughrul) Muḥammad *Beg* and Da'wūd Chaghri (Chekri, Jaqri),[111] both the sons of Mikhā'īl.[112] At that time, other Turkic tribesmen crossed the borders of the Abode of Islam. It looked like a massive *volkerwanderung* of Central Asian peoples. From 1035 the Saljūqs directly threatened the Iranian plateau.

The Saljūqs took advantage of the harsh climatic conditions.[113] While the mighty Ghaznavid army was paralysed, the three Saljūq brothers found shelter in neighbouring provinces. Chakri Beg defeated the Ghaznavid governor. Soon he was recognised as the independent ruler of Marv (428/June 1037). At this point in time, the Saljūqs sent an official letter (*nāma*) to the head of the Ghaznavid civilian administration (*ʿamīd*) of Khurāsān. This is probably the earliest Saljūq official document and it illuminates lucidly their self-representation in an early stage of their history:

> To his Lofty Highness, our Master Abū al-Faḍl Surī b. Muʿizz, from the slaves Bayghu, Ṭughril and Da'wūd [Chaghrī], humble servants of the Commander of the Faithful. We, the bondsmen, can no longer stay in Transoxania and Bukhara. That is so because ʿAlī Tegin [b. Bugha the Qarakhanid], who had good and friendly connections with us, died. The task fell to [his] two sons and the children are infants. Tunush, who had served as ʿAlī Tegin's army chief (*sipāh-salār*), controls them and controls the dynasty and army. He openly displayed enmity toward us till we were not able to live there. [Since] the killing of Hārūn [b. Altun-Tash, 1032–34] in Khwarazm, ceaseless troubles have befallen us and we cannot go thither. Because of these reasons, O Ruler of the World, the Great Sultan and the Most Merciful, protect us. We address you through this gentleman [al-Surī]. He promised to be of assistance and to write to the Chief Master, Aḥmad b. ʿAbd al-Ṣamad, asking him to serve as our advocate in the Sultan's court. He is familiar with our case. His intervention enabled us and our folks and animals to stay during wintertime and spring in the territory of the late Khwarazm-Shah in the days of the late Altun-Tash.
>
> We assisted our chief Lofty Master. Would he take us into his service? One of us will come to the Royal Sublime Court (*dargāh*) to serve there and the other will perform any sort of duty required. We are in great distress. The provinces of Nasa and Farawa that lie on the edge of the desert are dear to us. We would dwell there and find relief. No wrongdoer coming from Balkhan, Dehistan, Khwarazm and the Syr Darya regions will pass. We shall chase the Iraqi and Khwarazmian Turkmens. If our Master does not respond to us, God forbid, we do not know what will befall us. We do not have a place on earth.[114]

The correspondence did not prevent violence. The crucial clash between the Ghaznavids and the Saljūqs took place at Dāndanqān in early spring 1040. The Ghuzz nomads, commanded by the Saljūqs, routed the forces of Masʿūd of Ghazna and became the new rulers of Khurāsān and Northern Iran.[115] Following this victory, the Saljūqs wrote to the caliph in Baghdad. This dispatch (*maktūb*) visibly reflects the fact that after the decisive success at Dāndanqān the Saljūqs brought their ideology into line with the traditional Islamic political vocabulary. Not surprisingly, the letter opens with a quotation from the Qur'ān (Āl ʿImrān 3: 162) "There is no victory save with God." The verse is followed by the claim:

> We [i.e. the Saljūqs] have spread justice (*muʿādala*) among the people and we have kept away from the path of injustice (*bi-dād*) and oppression (*jawr*). We crave that state affairs will be in harmony with religion, in accordance with the laws of Islam (*qānūn-i islām*), and in fulfillment of the commands of the Commander of the Faithful.[116]

Ṭughril Beg was acknowledged at this point as the ruler of Ṭabaristān and Gurgan (Jūrjān). In the following year he was able to seize Khwarazm. These achievements ensured Ṭughril's and Chaghri's position as the chief Ghuzz leaders. From now onwards, the Arab chronicles grant more space to the two brothers' deeds and to the achievements of the Saljūqs.

Developments in Iraq facilitated the Saljūq advance towards Baghdad. The true master of the ʿAbbāsid capital was Abū al-Ḥārith al-Basāsīrī, a Turk general who commanded the *atrāk* guard, with the caliph al-Qāʾim no more than a puppet. The Saljūqs defeated him and conquered Baghdad in 1055. The caliph al-Qāʾim bestowed the title of *sulṭān* on Ṭughril Beg. In addition, the Saljūq leader was entitled *rukn al-dawla* (Pillar of the [ʿAbbāsid] Dynasty), *qasīm amīr al-muʾminīn* (Partner of the Commander of the Faithful), and *malik al-mashriq wal-maghrib* (King of East and West).[117]

Accomplishing these achievements, the Saljūq sultan became the de facto ruler, while the scion of the Prophet Muhammad's family was merely a symbolic figure. Ṭughril's ties with the ʿAbbāsid house were consolidated further when the caliph married Ṭughril's niece, Khadīja Khātūn in 1056.[118] Later, Ṭughril himself would marry one of al-Qāʾim's daughters. The Eurasian barbarian hoped to blend with the house of the Prophet.[119] In later Saljūqid writings, the Saljūq sultan was even portrayed as the restorer of religion.[120]

12 Ibn Ḥassūl: the structure and subject matter of his epistle

Turkic tribesmen were the backbone of the armies that carried the Saljūq family to Baghdad. This family and the *atrāk* are the heroes of the *Kitāb tafḍīl al-atrāk ʿalá sāʾir al-ajnād* ("The superiority of the Turks over the other troops"), an epistle written by Ibn Ḥassūl. The content of this treatise shows that the sultan and his secretaries were aware of the fact that a new chapter in the history of the Turkic people and the Caliphate had begun.

The data on Abū al-ʿAlāʾ Muḥammad b. Abi al-Qāsim ʿAlī b. Ḥusayn Ibn Ḥassūl is sparse,[121] although he was called in some sources "the sincere friend of the two royal courts" (*ṣafiyy*[u] *al-ḥaḍratayn*).[122] From the partial information available, we know that he was born in western Iran, in the town of Hamadhān, into a family of state secretaries. Moving to Rayy, then the capital of the province of Jibāl, he acquired a profound knowledge of Arabic, in addition to his native Persian tongue. Due to his linguistic talents, Ibn Ḥassūl obtained a position in the chancellery (*dīwān al-rasāʾil*) of the Ghaznavid dynasty.[123] A couple of years later, following the capture of Rayy by Ṭughril Beg (434/1042),[124] Ibn Ḥassūl became a clerk in the Saljūq administration.[125] He joined the new masters of northern Iran when they set out to occupy Iraq, where he passed away in 1058.

The composition of books (*muʾallafāt*) dealing with the history of the Saljūqs had already started during their initial tenure. The epistle by Muḥammad Ibn Ḥassūl was among the earliest contributions written under the Saljūqs, if not the first one. It is the only work from this early stage of Saljūqid historiography that has reached us. From the early decades of the Sultanate, not a single volume survived, although some quotations from now-lost books were incorporated into later writings.[126] It seems sound to conclude that the version of Ibn Ḥassūl's manuscript that is at our disposal contains only the introductory section of the original text.

To ask questions about the motives, aims and the nature of an author's narrative is an attempt to reflect on the culture and political discourse of his time and environment,[127] in our case on Ibn Ḥassūl and the early Saljūqid Sultanate. Although only a section of his work has reached us, these folios reveal the aims of the author. Hence, a short excursus on the narrative of the Turks' history is in line with the general theme of this introduction.

At the time Ibn Ḥassūl wrote his short epistle (*risāla*), he was serving as a Saljūq official. His epistle should be regarded as a political instrument. He praised the Turks in order to bolster the political claims of the Saljūq *sulṭān* Ṭughril Beg (d. 1063). Ibn Ḥassūl also eulogised ʿAmīd al-Mulk al-Kundurī (d. 1064), the chief vizier of the sultanate.[128] Ibn Ḥassūl's epistle provided his masters with a rationale that established their position as the supreme power in Iran and Iraq.

Being the servant of a Saljūq sultan and of his vizier, Ibn Ḥassūl endeavours to persuade his audience that the Saljūqid Sultanate is a legitimate political order and that the Arabo-Iranian secretarial class is accomplishing a most needed duty. The discourse employed by him is aimed at bringing the Turks into the heart of Islamic history. Bias and polemic drives motivated him to eulogise his sultan's people. He strives to demolish the old negative representation of the *atrāk*, as well as presenting the Saljūqs as the revivers of Sunnī tradition.

By the time Ibn Ḥassūl set about writing his glorifying epistle, the literary genres of ethnography and panegyric were deeply rooted in Arab culture. It seems that in the eleventh century, the late ʿAbbāsid period, these types of composition drew from three literary traditions that a learned man was supposed to be familiar with: 1) history; 2) ethnography; and 3) political discourse. A positive reception of the epistle could maintain the Saljūqs' claims to lofty position.

To achieve his purposes, Ibn Ḥassūl corresponds saliently with the *Kitāb al-Tāji*, which al-Ṣābī wrote for his Buwayhid master ʿAḍud al-Dawla. He chose deliberately to compose his book as a polemical treatise, using al-Ṣābī's *Kitāb al-tājī* (*The Book of the Crown*)[129] as a model, and simultaneously as an object of critique. On the one hand, Ibn Ḥassūl employed al-Ṣābī as his point of departure, structuring his composition in line with *al-Tājī*, yet on the other hand, he passed lengthy judgements on al-Ṣābī and on *Kitāb al-tājī*. Composing his narrative as a reaction to this lost work, Ibn Ḥassūl chooses to create a sharp antagonism between the Buwayhid past and the Saljūq present. He goes out of the way to remonstrate against al-Ṣābī and his history book, to condemn ʿAḍud al-Dawla, to denounce his people [130] and to elevate Ṭughril Beg and the Saljūq Turks. To uphold the case of their leader, Ṭughril Beg, Ibn Ḥassūl downplays the Buwayhids.

In his work, Ibn Ḥassūl shows great conformism. The position adopted by him is reflected in his treatment of ancient Persian mythology. Earlier Arab authors that set out to narrate the history of Iran dealt with that mythology and wrote a story that is a mixture of myth, epic, imagination and historical facts. Several works bear evidence that these authors collected data from ancient Persian sources.

In these Muslim chronicles, the memory of the royal fortune (*farr*) of the Iranian King of Kings was alive. Some authors even assumed that the Islamic dynasties were the legitimate successors of the Persian-Sasanian empire. They did not hesitate to depict the pre-Islamic past of Iran as a model that the Islamic administration should adopt, not only in the management of taxation and bureaucracy but also in the organisation of the Islamic religious establishment.[131]

Persian never ceased to be the lingua franca in the eastern parts of the Caliphate. Under the canopy of Islam, ancient myths were canonised in the plateau that is now the homeland of the Iranians. Narratives about the grandeur of ancient Iranian dynasties were used to reconstruct stories that Muslim writers transmitted during the ʿAbbāsid age. The revival of creative writing in Persian added energy to this literary tradition. This phase in Iranian literature culminated in the great Persian epic, the *Shāh-Nāma* (*nameh*; the Book of Kings) by Abū al-Qāsim Ferdowsī (also written Firdausi; Firdawsi; Firdusi; Ferdosi; *c*. 934–1020).[132]

In Ferdowsī's narrative, one encounters Turks, Indians and Arabs. Often the epic portrays their relations with Iranians as the struggle between the forces of light (good) and darkness (evil). According to this foundation myth, humanity was divided between the three sons of the legendary king Feridun: Salm, who represents the West; Tūr, who governed China and the Turks, that is, the East; and Irāj, who was seated in Iran and Iraq, the central parts of the world. Envy and competition between the brothers led the two elder sons to kill their younger brother, the ancestor of the Iranian people. The struggle between Iran and the nations surrounding its borders seems to be eternal. The Turks represent the calamity that the Iranian hero must defeat. They join forces with Ahriman, the evil god, against bright Feridun.

Focusing on ethnic images is also a salient feature of Ibn Ḥassūl's epistle. He continued the mytho-historical narrative of the Iranian past and built upon the ethnic discourse that prevailed in the ʿAbbāsid-period *adab* literature. Building upon this historical tradition and responding to the earlier Islamic account of the Central Asian peoples, Ibn Ḥassūl places the relationship between the Turkic peoples and the sedentary population of the Caliphate in a virtual framework of historic confrontation between Irān and the nomads of Central Asia (Tūrān).

At this junction we should mention the prominent place that Alexander the Great (Iskandar) occupies in Ibn Ḥassūl's epistle.[133] Starting from the early Islamic period, Alexander was commonly connected to the Qur'ānic *Dhū al-Qarnayn* (the Horned King; the king with two beams of glory or salvation) who is said to have confined the imaginary nations of Gog and Magog behind a barrier (*radm*).[134] Ṭughril Beg follows, according to Ibn Ḥassūl, in Alexander's footsteps.

In conjunction with the story of the Macedonian king, Ibn Ḥassūl also dwells on the story of Alexander's adviser (vizier; *wazīr*), the Greek philosopher Aristotle, who was viewed by Muslim scholars as a model for wisdom and learning.[135] In this matter Ibn Ḥassūl follows the Persian historical narrative of his day.

Ibn Ḥassūl took a difficult job upon himself and he faced great challenges. The aspiration of reshaping the popular image of the *atrāk* was not an easy task. His aims were to dismantle this image and to replace it with a constructive picture that would please his Saljūq masters. He had to contend with the negative representations of the Turkic nomads, who were portrayed as a menace that threatened the very existence of mankind. This drove him to dwell upon the mytho-history of Irān and Tūrān, as well as responding to earlier accounts of the Eurasian peoples.

Ibn Ḥassūl was determined to represent the success of the Saljūqs not as an accidental development, but rather as the culmination of a process. The competition between the Saljūqs and the Buwayhids (personified by 'Aḍud al-Dawla) was depicted as the zenith of a long historical development. It was not a temporal switch between emerging and descending forces. He argued that the roots of this success can be identified in Eurasian history. Ṭughril Beg was represented as a man of majestic origin.

The Saljūqs were said to have originated from a royal house, and their conversion to Islam did not result from coercion, but rather from voluntary acceptance of the true religion. They arrived in the central Islamic lands to support the declining 'Abbāsid caliphate and to revive Islam. In a society that afforded great importance to antecedence (*sābiqa*) and to kinship (*nasab*), Ibn Ḥassūl produced a fitting narrative for the Saljūqs' forefather. He provided him with bravery and his lineage with religious identity.

The Saljūqs were keen to establish their legitimacy on solid Islamic foundations and to avoid being portrayed as leaders of a barbarian incursion. Ibn Ḥassūl was determined to transmit his masters' ideology and his efforts served to gain them a new image. Among other things, the epistle "on the virtue of the Turks" doubtless provided the Saljūq sultan with an instrument to legitimise his rule.

13 The new image of the Turks

Images are deeply rooted and changes take place slowly. A hundred years after the conquest of Baghdad, in the decades that witnessed the gradual dismemberment of the Saljūq empire, the sources still praised the qualities of the *atrāk* as warriors and horsemen. Ibn Funduq (*c.* 1097–1169)[136] states that the people of each inhabited land (*vilāyeti*) have their particular qualities. The Turks are described by him as masters of mènage (*savarī*), skilled horsemen well-qualified in the usage of weapons.

The Saljūqs embraced the Sunnī dream of Islamic unity and ruled an empire stretching from the Mediterranean to the Inner Asian steppes. This might explain, at least partially, their readiness to assume the old Islamic world vision that was constructed by the echelon of Muslim-Iranian clerks that served their predecessors. The new regime inaugurated by Ṭughril Beg (d. 1063) did not wish to change the basic principles of the political discourse prevalent during the late ʿAbbāsid era, although the structure and administration of the Caliphate went through profound transformations.

Under the Saljūqs, the caliph's court in Baghdad lost its status. The title *sulṭān*, that in the past signified authority, became the royal title of the commander in chief of the armies and a new position, that of the sultan's *wazīr* (vizier), was created. He and other government officials, including the treasurer, accompanied the sultan during his travels. The sultan's court served also as the supreme court of appeal (*maẓālim*).[137]

Inspired by both Islamic-Iranian imperial and Turkish tribal traditions, the Saljūq sultans and their viziers wished to re-create a centralised government resting on three bases: 1) Sunnī legitimacy yielded by the ʿAbbāsid caliph; 2) a bureaucracy inspired by Persian tradition, which handled daily affairs, including diplomacy, taxation and land-tenure; and 3) an army composed of a small corps of paid retinues and the massive force of Ghuzz and other Turkic tribesmen. This was not only in accordance with the tradition of the Eurasian steppes but was also necessary for meeting the challenges of conquest and migration that faced the new Islamic regime in Western Asia.

The following tale illustrates this development clearly. Crossing the Euphrates River on his way from Edessa (Rūha; presently Orfa or Urfa in southern Turkey) to Syria, Alp Arslān, the Saljūq sultan of Iraq and Iran, was approached by a judge from Aleppo who told him:

> O our lord, thank God. Till now no free Turk has ever crossed this river, only Turk slaves (*mamlūk*) have done this so far. You crossed it as a king (*malik*). He [Alp Arslān] ordered the army commanders and the soldiers to come forward and instructed the *qāḍi* to repeat the saying.[138]

Several weeks after this incident, in August 1071, a fierce battle took place near Manzikert (Manazijird). It ended with a decisive victory by the army of Alp Arslān. This triumph was the first stage in the transformation of Byzantium into Turkey.[139]

The new order established by the Saljūqs was represented not only in historical writings or clerks' guidebooks, but also in legal-political works. Niẓām al-Mulk, the eminent vizier of the Saljūq Sultanate, characterises the Turks in his mirror for princes as "pure Muslims of the Ḥanafī sect". He argued, in response to critical voices, that the Turks originated from the Iranian world and were not barbarians from the steppes. They ousted the Daylamites-Buwayhids, who represented corruption and oppression, and were accused of being Shīʿīs.

Niẓām al-Mulk does not consider the ʿAbbāsid caliph of his day as the apex of the political and administrative structure. He claims that because of his failure to prevent unrest and bloodshed, the caliph's duties passed into the hands of the Saljūq sultan: "In every age and time God chooses one member of the human race and entrusts him with the interests of the world." The sultan is endowed with the prerequisites of power. However, since power alone is not enough, justice should be the foremost guiding principle of his authority.[140]

Abū Ḥāmid al-Ghazālī can unhesitatingly be considered one of the best-known Muslim thinkers. Born in Khurāsān in 1058, when Ṭughril Beg was at the peak of his career, he entered the Niẓāmiyya School (*madrasa*) in Nishapur. Later he moved to Baghdad and was appointed by the vizier Niẓām al-Mulk to the post of teacher at the Niẓāmiyya *madrasa* in Baghdad, the institution of higher learning that this able administrator had established in 1091.

In al-Ghazālī's writings, the caliph (*khalīfa, imām*) symbolises the unity of the community of believers, and he is the keeper of the harmony of religion. He does not hold executive power, but delegates it to an associate, the ruler (*sulṭān*), who has to be supported by a skilful vizier (*wazīr*). The *imām* and the *sulṭān* both share the burden of power.[141] This political construction echoes the framework of Ibn Ḥassūl's vision of Ṭughril Beg's universal rule.

It was already said that Ibn Ḥassūl's text reflects the official version of the Saljūqs' historiography. The great Saljūqs became the de facto rulers of Western Asia. Historians who worked at their courts narrated the history and foundation accounts of the ruling dynasty. In the generations that followed, other Arab and Persian writers set out to narrate the story of the Turks and the Ghuzz. This literary interest in Eurasia and the Turkic peoples resulted from the new position of the *atrāk* within the Abode of Islam. They acquired further praise due to the *atrāk*'s role in fighting the Franks and the Mongols. This certainly contributed to the new representation of the *atrāk* in late medieval Arabic historiography.

One example that should be mentioned here is a book by Abū al-ʿAbbās Aḥmad b. ʿAlī Ibn Bābah (or Bābā) al-Qāshī (or al-Qāshānī), an Iranian author (*adīb*) (d. 1166–67), who produced several works on *ḥadith*, *adab* and *ta'rīkh* (history).[142] One of his *adab* works, labelled *Rās māl al-nadīm fī tawārīkh aʿayān ahl al-islām* ("The boon companion's stock in trade: about the history of noble Muslims", composed in 1107–08), presents the history of Islam. This book contains some valuable information on historical events that took place on the Iranian plateau. According to Ibn Bābā, he dedicated his work to the history of the Saljūqs, their emergence, victories, conquests and the new government that they established in Khurāsān, Iraq, Byzantium and Syria.[143]

During Ibn Bābā's time, an anonymous Iranian author composed a collection of historical accounts and stories. It is assumed that the *Mujmal al-tavārīkh wal-qiṣaṣ* was compiled around 1126.[144] The book includes a concise account of Ṭughril Beg's history, but not a word about his family's origin:

> The banners of the grand sultan Abū Ṭālib Muḥammad b. Saljūq, who bore the honorific title (*laqab*) Ṭughril Beg, were exhibited in Khurāsān, in the northeastern [districts of the Iranian plateau]. This event opened the reign of the [Saljūqid] dynasty (*dawla*). He defeated the [Ghaznavid] sultan Masʿūd

b. Maḥmūd in the year 431/1040 [in Dāndanqān]. Consequently, following a truce he agreed upon with [Abū Manṣūr Farāmarz] Ibn ʿAlāʾ al-Dawla (433–443/1041–51),[145] Ṭughril Beg seized Rayy and Isfahan. He assigned Abū Manṣūr Farāmarz the tax farming (*iqṭāʿ*) of the district of Yazd and Abarquh. Ibn ʿAlāʾ al-Dawla settled there. ʿAmīd al-Mulk Abū al-Naṣr al-Kundarī was nominated vizier.

Ṭughril proceeded to Baghdad, liberated the caliph from al-Bāsasīrī's cage and reseated him on the caliphs' throne. Ṭughril advanced towards the borders of Syria and Byzantium. He gained great victories and conquered vast lands. Ṭurghil Beg hastened from Syria and speedily, in less than two weeks, arrived at Hamadhān with several horseriders. He acted so because (literally: at the time) his brother Sayf al-Dawla Ibrāhīm Yināl (Īnāl) besieged the city.[146] [Ṭurghil Beg] reached [the scene] and Sayf al-Dawla was killed. As a result his reign was the most blessed and no dynasty compared to the bliss of his days. His excellent mode and character remained in this very manner among all the offspring of the Saljūqid house. Ṭughril Beg passed away in the city of Rayy in the year 456/1064 (sic). He served as sultan during the term in office of the caliph al-Qāʾim. His reign lasted 26 years. After him Alp Arslān became the great sultan.[147]

Later Iranian pro-Saljūqid sources make efforts to represent the Saljūqs as nobility whose roots can be traced back to an ancient Iranian father or a mythological Arabian ancestor. While some would argue that the Saljūqs are the offspring of Afrasiyab (Afrasiab), others claim that Lūqmān, a pre-Islamic wise man mentioned in the Qurʾān, was Saljūq's ancestor. Saljūq the son of Duqāq is said to have served as an army commander (*sū-bāshi*) of the Turkic king in the steppes. He is described as an audacious man, blessed with renown, who had four sons, all called by biblical names: Bayghū Mūsa (Sparrowhawk Moses), Mikhāʾīl (Michael), Arslān Isrāʾīl (Lion Israel) and Yūnis (Jonas). These sons are described as charismatic leaders. Each day their followers and companions were on the ascendant and increased.[148]

The accounts presented in the last paragraphs seem to echo official Saljūqid propaganda. This deduction rests upon the resemblance between the historical accounts and a letter (*fatḥ nāmah*) dispatched by the Saljūqs to the caliph al-Qāʾim in 1039.[149]

14 From al-Makīn ibn al-ʿAmīd to Mamluk historiography

Nevertheless, the favourable representation of the Saljūqs was not limited to historians operating in their domains. Interest in Saljūqid history did not evaporate in the post-Saljūqid period. The dynasties that emerged in Syria drew their legitimacy from the Saljūqs. The Ayyūbid sultanate witnessed the flourishing of historical writing. Several Syrian and Egyptian authors incorporated chapters on the history of the Turks in their writings. These chroniclers narrated the history of the Saljūq sultanate and its disintegration.

The best known of these Ayyūbid-period chroniclers is ʿIzz al-Dīn ʿAlī b. Abī al-Karm Muḥammad Ibn al-Athīr al-Jazīrī (1160–1233). He certainly can be seen as the chief contributor to the historical library of those years. His *al-Kāmil* (*The Complete History*) is a major source for the study of Saljūqid history,[150] particularly because of his use of earlier sources, including the *Malik nāmah* and *Zubdat al-tawārīkh*.[151]

An additional illustration of this curiosity about the Saljūq past can be traced in the chronicle of the great Syriac Christian scholar Gregory Bar Hebraeus, who lists among his sources on the Saljūq kings a Persian book, which he calls "The Book of Royalty (*mūlk nāmah*)".[152]

Almost a generation after Ibn al-Athīr's death, Jirjis ibn al-ʿAmīd, known as al-Makīn, a Christian subject of the late Ayyūbids (d. 1273),[153] compiled his chronicle in 1260. In his condensed world history, named *al-Majmūʿ al-mubārak* (*fī al-taʾrīkh*),[154] al-Makīn provides a chronology of Islam from the days of the Prophet Muḥammad until the days of the Mamlūk sultan al-Ẓāhir Baybars.[155]

The chapter on "The Emergence of the Saljūq Dynasty" (*ibtidāʾ dawlat al-salāṭīn al-saljūqiyya*) seems to match the work of Jamāl al-Dīn Muḥammad b. Sālim Ibn Wāṣil (d. 1298).[156] Since both authors wrote at the same time, more or less, it is difficult to tell who wrote the original and who borrowed. Moreover, we should not rule out the likelihood that both depended on a common source.[157] It is also difficult to accept the claim that al-Makīn summarised Ibn al-Athīr's *al-Kāmil*, since despite the fact that several sentences in al-Makīn's chronology correspond largely to Ibn al-Athīr's world history, long paragraphs are considerably different.

The sequence of events that Jirjis ibn al-ʿAmīd al-Makīn's short chapter on the "Beginning of the Saljūq Dynasty" puts forward is that the Saljūqs converted to Islam willingly in the Eurasian steppes, years before they moved to the Abode of Islam. The ancestors of the family fought for the cause of Islam and participated in the *jihād* against the infidel (*kuffār*) Turks. They entered the Abode of Islam in response to an invitation by the legitimate rulers, and the ʿAbbāsid caliph requested them to protect his territories and encouraged them to fight the enemies of Islam.

Despite the fact that the historians described here do not keep silence on military defeats of the Saljūqs and report on atrocities committed by them, their texts reflect the official (late) Saljūqid argumentation to a considerable degree. These annals reproduce a conventional account of the chronology of the Saljūqid sultanate. They demonstrate that the official propaganda of the Saljūqs was received by successive generations and became embedded in popular historical writings. Muslim and non-Muslim authors recycled this story of the past as the standard historical narrative of the eleventh-century Near East.

The pro-Saljūqid version of history is clearly visible in chronicles composed during the years of *jihād* against the Franks, in which *atrāk* armies played a significant role. It seems also that the abovementioned historians regarded chronicles as a useful tool. They maintained that for utterly practical reasons, kings and governors should acquire a good knowledge of past events and fighting.[158]

The slow process of *atrāk* and other Eurasians' acculturation within the Abode of Islam reached its apex with the emergence of the Mamlūk Sultanate. Writers who were working in the service of the Mamlūk rulers depicted their masters as the force that saved Islam from extinction.[159] They represented them as courageous fighters who sacrificed their lives in fighting against the Franks and the Mongols.

Any attempt to narrate the history of Mamlukdom (*dawlat al-atrāk*) would divert us from this study's main path. Hence we will limit ourselves to saying that although the contemporary historians pictured the Mamlūks of Egypt and Syria as manumitted slaves (*massahum al-riqq*), no contemporary author doubted the Mamlūk regime's legitimacy. Moreover, religious functionaries and clerks competed to be recruited by sultans and officers.[160] Authors who commanded Turkish or were born into Turkic households used their qualifications to attract the attention of high-ranking commanders and officials. The historical discourse spun by them bears no traces of hostility towards the Eurasian people.

The historical works of Abū Bakr b. ʿAbd al-Allah b. Aybak Ibn al-Dawādārī (*c.* 1288–1336) betray a taste for "the strange and wonderful". His minor chronicle, the *Durar al-tījān*,[161] and his major chronicle, the *Kanz al-durar*,[162] which narrates the history of Egypt, are broken up by records of *mirabilia* and by excurses on Turkish legend and folklore. He defended the Turks against accusations of superstition and presented Baybars as a ruler in the tradition of the Saljūq sultans.[163] One of the Turkish stories that Ibn al-Dawādārī brings seems to be extracted from an early version of the Turkish epic *The Book of Dede Korkut*.[164] This is a creation legend, which differs from the biblical-Islamic one.

Badr al-Dīn Maḥmūd b. Aḥmad al-ʿAynī (1360–1451) was a Turk and, perhaps because of his knowledge of Turkish, was successful in attracting the friendship and favour of the Mamlūk sultans. Two of his chronicles, *al-Sayf al-muhannad fī taʾrīkh al-malik al-Muʾayyad* and *al-Rawd al-zāhir fi sīrat al-malik al-Ẓāhir Ṭaṭar*, have a quotidian political character; both works aim at bolstering the legitimacy of the sultans.[165] In both treatises al-ʿAynī provides a miscellaneous range of justifications for their rule, presenting them as ideal Turkish sultans.[166] The chapter on the early history of the Turkic people and their tribal organisation is a central element in an account that aims to provide manumitted slaves with a glorious, if remote, past.

Using early sources, al-ʿAynī provides data for the pre-Islamic chapter in the history of the Eurasian steppes found in earlier Arabic chronicles,[167] including the brands (*tamghā*) used by the Oghuz tribes. In addition, al-ʿAynī's account reflects the internalisation of the Islamic history of mankind by the Turks. In these accounts Alexander acquires a new role. The Macedonian king, who is said to imprison Gog and Magog, the mythological ancestors of the Eurasian people, behind a wall, is depicted as the link between the Turks and mankind. In these accounts, the Turks are not presented as a menace, but rather as the Saviours of Islam.

In conclusion

The coming of the Turks and the emergence of the Saljūq Sultanate opened a new chapter in the history of Iran, Western Asia and Islam. The mass migration of

Turkic tribes had demographic implications and reshaped society in the lands they reached.[168] The acquisition of the actual seats of government in large parts of the Abode of Islam by the *atrāk* and the transition of rulership from caliphate to Saljūq sultans reflects a deep change in Islamic political discourse. The third dimension of change was the place occupied by tribal armies and regimes that eradicated ideological governments.[169] The historiography of these developments was not detached from the changes in political theory that developed during the second half of the Islamic Middle Period. The polity and the discourse both witnessed changes and amendments; however, this topic is beyond the scope of this short introduction.

Notes

1 Denis Sinor (ed.), *The Cambridge History of Early Inner Asia* (Cambridge, 1990).

2 *Chronicon pseudo-Dionysianum*, ed. Jean Baptiste Chabot (Louvain, 1952), 2: 168 (ll. 13–15); Amir Harrak (trans.), *The Chronicle of Zuqnin: part 3&4 A.D. 488–775* (Toronto, 1999), 159; Johan Malalas, *The Chronicle*, trans. E. Jeffreys et al. (Melbourne, 1986), 249; Procopius of Caesarea, *History of the Wars*, ed. and trans. H.B. Dewing (London, 1961), book 1, 3, 1; Theophylact Simocatta, *The History of Theophylact Simocatta*, trans. M. and M. Whitby (Oxford, 1997), book 7: 6–9; *The Armenian History Attributed to Sebeos*, trans. R.W. Thomson, J. Howard-Johnston and Tim Greenwood (Liverpool, 1999), 14–15, 184–8.

3 The word "Turk" is mentioned in several pre-Islamic verses. In an ode (*qaṣīda*) attributed to Abū Ṭālib in A. Guillaume (trans.), *The Life of Muhammad* (Oxford, 1955), 124 (l. 6), it seems to hint at a location or mountain in southern Syria; see T. Kowalski, "Die aeltesten Erwaehnungen der Tuerken in der arabischen Literatur", *Koeroesi Csoma-Archivum* 2 (Budapest, 1932): 38–41.

4 For a survey of the various opinions over this people, see M.D. Dunlop, *The History of the Jewish Khazars* (Princeton, 1954), 3–33; Peter B. Golden, *Khazar Studies: A Historico-philological Inquiry into the Origins of the Khazars* (Budapest, 1980), 1: 51–7, 262–3.

5 Although several Arabic sources cast light on the history of this vast region, I will limit the reference to these texts. Sources for further reading are included in the Bibliography.

6 A fresco in the Umayyad Quṣayr ʿAmra in Jordan depicts a woman said to be the queen (*khātūn*) of Bukhara. P. Wittek, "Eine Turkische Furstin auf dem Wandgemalde von Kusajr ʿAmra", *Der Islam* 14 (1924): 402–6; H.A.R. Gibb, "Kleine Mitteilungen und Anzeigen zu *Der Islam* XIV 402—406", *Der Islam* 20 (1932): 196–7; E. Herzfeld, "Die Könige der Erde", *Der Islam* 21 (1933): 233–6; O. Grabar, "The Painting of the Six Kings at Qusayr ʿAmrah", *Ars Orientalis* 1 (1955): 185–7.

7 A.D. Mordtmann (trans.), *Schech Ebu Ishak el Farsi el Isztachri – das Buch der Länder* (Hamburg, 1845), 104–5.

8 Mohsen Zakeri, *Sassanid Soldiers in Early Muslim Society: The Origin of ʿAyyaran and Futuwwa* (Wiesbadan, 1995), 57–68, 112–15.

9 D.G. Tor, "The Mamluks in the Military of the Pre-Seljuq Persianate Dynasties", *Iran* 46 (2008): 213–25, argues against the prevailing interpretation of this phenomenon. A close analysis of her thesis is beyond the limits of the present work; however, we can point out that Turkish fighters were also recruited by Byzantine emperors. See Charles M. Brand, "The Turkish Element in Byzantium, Eleventh-Twelfth Centuries", *Dumbarton Oaks Papers* 43 (1989): 1–3; and cf. Edwin G. Pulleyblank, "A Sogdian Colony in Inner Mongolia", *T'oung Pao*, second series 41–4/5 (1952): 349.

10 Issac Hasson, *Recherches sur Muawiya Ibn Abi Sufiyan* (Ph.D. dissertation, The Hebrew University of Jerusalem, 1982; in Hebrew), 295–328; M. Sharon, *Revolt: The Social and Military Aspects of the ʿAbbāsid Revolution* (Jerusalem, 1990), 272–3.

11 R. Şeşen, *Hilafet ordusunun menkıbeleri ve Türklerin fazailetleri* (Ankara, 1967), 28–9; Patricia Crone, *Slaves on Horses: The Evolution of the Islamic Polity* (Cambridge, 1980); Daniel Pipes, *Slave Soldiers and Islam: The Genesis of a Military System* (New Haven, 1981); Hugh Kennedy, *The Armies of the Caliphs: Military and Society in the Early Islamic State* (London and New York, 2001).

12 P. Crone, "The Significance of Wooden Weapons in al-Mukhtar's revolt and the Abbāsid revolution", in I.R. Netton (ed.), *Studies in Honour of Clifford Edmond Bosworth* (Leiden, 2000), 1: 175–6, and the critique by Amikam Elad, "The Ethnic Composition of the Abbāsid Revolution: A Re-evaluation of Some Recent Research", *Jerusalem Studies in Arabic and Islam* 24 (2000): 275–82.

13 Cf. the list of names in Amabe Fukuzo, *The Emergence of the Abbāsid Autocracy: The Abbāsid Army, Khurasan and Adharbayjan* (Kyoto, 1995), 141–4.

14 Osman S. A. Ismail, "Muʿtasim and the Turks", *Bulletin of the School of Oriental and African Studies* 29 (1966): 12–24; David Ayalon, "The Military Reforms of Caliph al-Muʿtaṣim: Their Background and Consequences", in his *Islam and the Abode of War: Military Slaves and Islamic Adversaries* (Variorum, 1994), art. 1: 28–9; Uthman Sayyid Aḥmad Ismaʾil al-Bili, *Prelude to the Generals: A Study of Some Aspects of the Reign of the Eighth ʿAbbāsid Caliph, Al-Muʿtasim Bi-Allah (218–277AH/833–842AD)* (Reading, 2001).

15 Osman S.A. Ismail, "The Founding of a New Capital: Samarra", *BSOAS* 31 (1968): 3–4.

16 Matthew S. Gordon, "The Khaqanid Families of the Early ʿAbbāsid Period", *Journal of the American Oriental Society* 121 (2001): 237.

17 Peter B. Golden, "Wolves, Dogs and Qipcaq religion", *Acta Orientalia* 50 (1977): 94 (reprinted in his *Nomads and their Neighbours in the Russian Steppe* (Variorum, 2003)).

18 Tarif Khalidi, *Arabic Historical Thought in the Classical Period* (Cambridge, 1994), 116–17.

19 David Ayalon, "The Military Reforms of the Caliph al-Muʿtasim – Their Background and Consequences" (reprinted in his *Islam and the Abode of War: Military Slaves and Islamic Adversaries* (Aldershot, 1994), 1–39); *idem*, "Preliminary Remarks on the Mamluk Military Institution in Islam", in V.J. Parry and M.E. Yapp (eds.), *War, Technology and Society in the Middle East* (Oxford, 1975), 53–5; Osman S.A. Ismail, "The Founding of a New Capital: Samarra", *BSOAS* 31 (1968): 1–13.

20 E.A. Wallis Budge (ed. and trans.), *The Book of Governors: The Historia Monastica of Thomas Bishop of Marga A.D. 840* (London, 1893), 1: 141 (l. 17), 2: 290 (translation).

21 G. Wiet (trans.), *Yaʿkubi – Les Pays* (Cairo, 1937), 45–55.

22 Matthew S. Gordon, *The Breaking of a Thousand Swords: A History of the Turkish Community of Samarra (200–275/815–889)* (Albany, 2001).

23 Nizar F. Hermes, *The [European] Other in Medieval Arabic Literature and Culture* (New York and Basingstoke, Hampshire, 2012) provides several remarks on the land of the Türk (taken mostly from Ibn Faḍlān's travelogue).

24 Aloys Sprenger (trans.), *el-Masudi's Historical Encyclopaedia* (London, 1841), 30, 237–45, 309–12. This important work was re-edited and re-translated into French by Ch. Pellat (trans.), *Les Prairies d'or* (Paris, 1962–1997), chapter 15; on this section, see Ahmad M. H. Shboul, *Al-Masʿudi and his World: A Muslim Humanist and his Interest in Non-Muslims* (London, 1979), 163–72.

25 Richard Nelson Frye, "A New Arabic Geographical Manuscript", *Journal of Near Eastern Studies* 8 (1949): 92–3 (a translation of chapters from the eleventh-century Arab-Spanish author Isḥāq b. Ḥusayn).

26 Andrew C.S. Peacock, *Early Seljūq History* (London, 2010), 18.
27 Suliman Bashear, *Arabs and Others in Early Islam* (Princeton, 1997), 104.
28 S.M. Stern, "The Arabic Translation of the Pseudo-Aristotelian Treatise *De Mundo*", *Le Museon* 77 (1964): 197–9, 202; *idem*, "A Third Arabic Translation of the Pseudo-Aristotelian Treatise *De Mundo*", *Le Museon* 78 (1965): 382–5, 388–9; M. Maroth, "The Correspondence between Aristotle and Aexander the Great: An Anonymous Greek Novel in Letters in Arabic Translation", *Acta Antiqua Hungaricae* 45 (2005): 245.
29 The *Secret of Secrets* is obviously not an Aristotelian work. Steven J. Williams, *The Secret of Secrets: The Scholarly Career of a Pseudo-Aristotelian Text in the Latin Middle Ages* (Ann Arbor, 2003), 7–17.
30 Mario Grignaschi, "Les *rasail Aristatalisa ila Iskandar*", *Bulletin d'Etudes Orientales* 19 (1965): 54–5. Some scholars doubt the Umayyad dating of this compilation.
31 Another *ḥadīth* says: "My offspring will hold the Caliphate until the red faces [i.e. the *atrāk*] seize their empire. Their faces resemble iron shields." Sulaymān b. Ash'ath al-Azdī Abū Dā'ūd al-Sijistānī (818–888), *Sunan Abī Dā'ūd*, ed. Muhammad M. 'Abd al-Hamid (Cairo, 1951), 4: 160 (*Kitāb al-malāḥim* traditions 4302–5); al-Muṭahhar b. Ṭāhir al-Maqdisī (fl. *c.* 960), *al-Bad' wal-ta'rīkh* (Paris, 1899), 2: 171. 'Abd Allah b. al-'Abbās has been reported to transmit the next tradition: "The hour of the resurrection shall not come until a nation with wide faces, narrow little eyes and flat noses will set up their houses on the shores of the Tigris River." The transmission of a similar tradition is attributed to Abū Hurayra. It is worth mentioning that this saying was in circulation, although with minor amendments, many years after the emergence of the Saljūq Sultanate. One of them says: "It is said that the 'Abbāsid caliphate will be destroyed by the Muslim Turks, and in their turn this people will be annihilated by the infidel Turks." Sirāj al-Dīn Ibn al-Wardī (d. 1457), *Kharīdat al-ajā'ib wa-farīḍat al-gharā'ib*, ed. M. Fakhuri (Aleppo, 1411/1991), 293. This negative vision of the Turks can be traced in Iranian writings: H.W. Bailey, "To the Zamasp Namak", *BSOAS* 6 (1930): 581 (l. 73).
32 Jalāl al-Dīn Abū al-Faraj 'Abd al-Raḥmān b. 'Alī Ibn al-Jawzī (1117–1201), *Akhbār al-ḥamqá wal-mughaffilīn*, ed. R.J. Radwan (Mecca, 1425/2004), 115 (*akhbār al-a'imma*).
33 Muḥammad b. Aḥmad b. Tamīm al-Tamīmī (d. 944), *Kitāb al-miḥan*, ed. Y. al-Jaburi (Beirut, 1408/1988), 147.
34 Abū 'Alī al-Muḥassin b. 'Alī al-Tanūkhī (939–995), *Nishwār al-muḥāḍara wa-akhbār al-mudhākara* (The Table Talk of a Mesopotamian Judge), ed. A. al-Shaliji (Beirut, 1391/1971–73), 150–4.
35 B. Lewis, *Race and Slavery in the Middle East: A Historical Enquiry* (Oxford, 1990), doc. 1, 143; *idem*, *Islam from the Prophet Muhammad to the Capture of Constantinople Islam* (New York, 1974), 2: 207–8.
36 G. Newby, *The Making of the Last Prophet: A Reconstruction of the Earliest Biography of Muḥammad* (Charlotteville, 1999), 199.
37 Gregorii Barhebraei (1226–1286), *Chronicon Syriacum* (Paris, 1890), 217 (ll. 16–17; Syriac); E. Wallis Budge (trans.), *The Chronography of Gregory Abū al-Faraj Bar Hebraeus* (Oxford, 1932), 195.
38 Bashear, *Arabs and Others in Early Islam*, 71.
39 One topic in the story of the great Hellenistic king is his relation with the Iranians. This topic was later taken up by several writers, including the well-known Persian poet Firdūsī. Th. Noldeke, "Beitrage zur Geschichte des Alexander-Romans", *Osterreichische Akademie der Wissenschaften, Denkschriften der Philosophisch Historische Klasse* CL. 38. Bd. 5 (Wien, 1890), 34–50; A. Abel, *Le Roman d'Alexander – Legendaire medieval* (Brussels, 1955), 9–13; Grignaschi, "Les *rasail* Aristatalisa ila Iskandar", 7–15; T. Nagel, *Alexander der Grosse in der fruhislamischen Volksliteratur* (Walldorf-Hessen, 1978), 149–52.

40 E.A. Wallis Budge (trans.), *The History of Alexander the Great, Being the Syriac Version edited from Five Manuscripts of the Pseudo-Callisthenes with English Translation* (Cambridge, 1889), 176–8; *idem* (trans.), *The Alexander Book in Ethiopia* (London, 1933), 226; Reuben Levy (trans.), *The Epic of the Kings: Shah-Nama, the National Epic of Persia by Ferdowsi* (Chicago, 1967), 146–8; al-Kāshgharī, *Dīwān lughāt al-turk* 3: 304–7; Dankoff, *Compendium of the Turkic Dialects*, 2: 362–3; A.R. Anderson, *Alexander's Gate, Gog and Magog, and the Enclosed Nations* (Cambridge, MA, 1932).

41 Qur'ān, The Cave, 18: 83–98.

42 William Ouseley, *The Oriental Geography of Ebn Haukal, An Arabian Traveller of the Tenth Century* (London, 1800), 158.

43 Faustina Doufikae-Aerts, *Alexander Magnus Arabicus: A Survey of the Alexander Tradition Through Seven Centuries: from Pseudo-Callisthenes to Suri* (Leuven, 2010), 20, 31, 83.

44 Minoo S. Southgate, "Portrait of Alexander in Persian Alexander-Romances of the Islamic Era", *JAOS* 97/3 (1977): 283.

45 Ehsan Yarshater (ed.), *The Cambridge History of Iran* vol. 3/1: *The Seleucid, Parthian and Sasanian Periods* (Cambridge, 1983), 377.

46 Emel Esin, "Tarkhan Nizak or Tarkhan Tirek? Enquiry Concerning the Prince of Badhghis Who in A.H. 9 1/709–710 Opposed the 'Omayyad Conquest of Central Asia", *JAOS* 97 (1977): 323–5.

47 In the printed edition of Qudāma it says: "Since I learned that all your victories result from the Almighty's wish."

48 Compare Qur'ān, 2: 249 "How many a little company hath overcome a mighty host by Allah's leave!" (trans. Pickthall).

49 Vladimir Minorsky, "Tamim ibn Bahr's Journey to the Uyghurs", *BSOAS* 12 (1948): 283 (item 7); R.B. Serjeant, "Material for a History of Islamic Textiles up to the Mongol Conquest", *Ars Islamica* 15–16 (1951): 82; Ch. Pellat, "Ǧāhiẓiana, I: Le *Kitāb al-Tabaṣṣur bi-l-tiğāra* attribué à Ǧāḥiẓ", *Arabica* 1/2 (1954): 149 (item 14: merchandise from China) erroneously translated as figured cloth/aromatics. For the interpretation "swords made from watered or moiré steel", see Afif Bahnassi, "Fabrication des épées de Damas", *Syria* 53 (1976): 287–88, 293; D.G. Alexander, "Two Aspects of Islamic Arms and Armor", *Metropolitan Museum Journal* 18 (1983): 104 (no. 1); Yūsuf Rāġib and Philippe Fluzin, "La fabrication des lames damassées en Orient", *Journal of the Economic and Social History of the Orient* 40/1 (1997): 36, 69.

50 John Andrew Boyle, "Alexander and the Mongols", *JRAS* 2 (1979): 129.

51 Qur'ān 18: 83–99.

52 John Andrew Boyle, "The Alexander Romance in the East and West", *Bulletin of the John Rylands University Library of Manchester* 60 (1977): 18–19.

53 Ibn Khurdadhbeh [Ibn Khurradādhbih], 'Ubayd Allah bn 'Abd Allah (lived *c.* 205-317/820-916). *al-Masālik wal-mamālik*, ed. M. De Goeje (Leiden, 1889; reprinted 1967), 261–3.

54 Dunlop, *The History of the Jewish Khazars*, 190–3; Andre Miquel, *La Geographie humaine du monde musulman jusqu'au milieu du 11e siecle* 2 – *Geographie arabe et represetation du monde: la terre et l'etranger* (Paris, 1975), 497–511.

55 For an early English translation of this text, see C.E. Wilson, "The Wall of Alexander against Gog and Magog and the Expedition sent out to find it by the Khalif al-Wathiq in 842 A.D.", in *Asia Major Hirth Anniversary Volume* (London, 1923), 575–612. This account was studied and translated: see Emeri van Donzel, Andrea Schmidt with a contribution by Claudia Ott, *Gog and Magog in Early Eastern Christian Syriac and Islamic Sources: Sallam's Quest for Alexander's Wall* (Leiden, 2010). Travis Zadeh, *Mapping Frontiers across Medieval Islam: Geography, Translation, and the 'Abbāsid Empire* (London and New York, 2011) deals with the caliph al-Wāthiq's translators and the journey of Sallām al-Tarjumān to the wondrous barrier.

56 S.M. Stern, "Ya'qūb the Coppersmith and Persian National Sentiment", in C.E. Bosworth (ed.), *Iran and Islam: Studies in Memory of the late V. Minorsky* (Edinburgh, 1971), 535–55.

57 Ruth Meserve, "The Inhospitable Land of the Barbarian", *Journal of Asian History* 16 (1982): 69–75; Y. Frenkel, "The Turks of the Eurasian Steppes in Arabic Geographical Literature", in M. Biran and R. Amitai (eds.), *Mongols, Turks and Others: Eurasian Nomads and the Sedentary World* (Leiden, 2005), 201–41.

58 The name of this social and literary movement goes back to the Qur'ān, al-Hujarat, 49: 13 "O mankind, We have created you male and female, and have made you races and tribes"; Ignaz Goldziher, *Muslim Studies*, ed. and trans. S.M. Stern (London, 1967), chs. 4 and 5; Hamilton A.R. Gibb, "The Social Significance of the Shu'ubiya" (reprinted in his *Studies on the Civilization of Islam* (London, 1962)); Lutz Richter-Bernburg, "Linguistic Shu'ūbīya and Early Neo-Persian Prose", *JAOS* 94/1 (1974): 55–64; Roy Mottahedeh, "The Shu'ubiyah Controversy and the Social History of Early Islamic Iran", *International Journal of Middle East Studies* 7 (1976): 161–82; H.T. Norris, "Shu'ubiyya in Arabic Literature", in J. Ashtiany et al. (eds.), *The Cambridge History of Arab Literature: 'Abbāsid Belles-Lettres* (Cambridge, 1990), 31–47.

59 al-Jāḥiẓ, "*Risala ila al-Fatḥ b. al-Khāqan fī manāqib al-atrāk*", ed. G. Van Vloten, *Tria Opuscula Auctore Abū Othman Abr ibn Bahr al-Djahiz* (Leiden, 1903), 4 (l. 15)–6 (l. 10); this epistle was translated by William M. Hutchins in *Nine Essays of al-Jāḥiẓ* (New York, 1975).

60 Lewis, *Race and Slavery*, appendix, doc. 1, 144.

61 Kai Ka'ūs Ibn Iskandar Ibn Qabūs (d. 1072), *A Mirror for Princes: The Qabous Nameh* trans. R. Levy (London, 1951), 230.

62 Pilar Coello, "Las Actividades de las esclavas segun Ibn Butlan (s. 11) y al-Saqati (ss. 12–13) de Malaga", in Maria J. Viguera (ed.), *La Mujer en al-Andalus: Reflejos historicos de su actividad y categories socials* (Madrid, 1989), 207.

63 Peter Malcolm Holt, "The Exalted Lineage of Ridwan Bey: Some Observations on a Seventeenth-Century Mamluk Genealogy", *BSOAS* 22 (1959): 221–30.

64 Stefan Leder and Hilary Kilpatrick, "Classical Arabic Prose Literature: A Researchers' Sketch Map", *Journal of Arabic Literature* 23/1 (1992): 11–14; Nadia Maria El Cheikh, "Revisiting the Abbasid Harems", *Journal of Middle East Women's Studies* 1/3 (2005): 2.

65 al-Tanūkhī, *Nishwār al-muḥāḍara*, 7: 904–5.

66 Zacharias van Laer, "Ibn al-Faqīh et le Livre des pays", *Arabica* 31 (1984): 69–79.

67 The version of *Kitāb al-Buldān* that has reached us seems to be incomplete. The abridgement copied by al-Shayzarī in 1022 was published by De Goeje in 1885. This manuscript does not include the chapter on the Turks. The Mashhad manuscript is another abridgement and does contain it. See O. Tskitishvili, "Concerning the Tashkent manuscript of a geographical treatise by Ibn al-Faqīh al-Hamadani", *JESHO* 24 (1980): 304–8.

68 A.B. Khalidov, "Myth and Reality in the *Kitāb akhbār al-buldān* by Ibn al-Faqīh", in Angelika Neuwirth et al. (eds.), *Myths, Historical Archetypes and Symbolic Figures in Arabic Literature: Towards a New Hermeneutic Approach* (Beirut, 1999), 481–2.

69 Vladimir Minorsky, "Tamim ibn Bahr's Journey to the Uyghurs", *BSOAS* 12 (1948): 275–305.

70 The journal was first published by Zegi Validi Togan, *Ibn Fadlan's Reisebericht* (Leipzig, 1939; repr., 1966). Sami Dahhan re-edited it as *Risālat Ibn Faḍlān* (Damascus, 1959). See Richard Nelson Frye with R. Blake, "Notes on the Risala of Ibn Fadlan", *Bizantina Metabyzantina* 1 (1949): 7–37; Marius Canard, "La relation du voyage d'Ibn Fadlan chez les Bulgares de la Volga", *Annales de l'Institut d'Etudes Orientales de la Faculte des Lettres d'Alger* 16 (1958): 41–146; T. Nagrodzka-Majchrzyk, "Les Oghouz dans la relation d'Aḥmad Ibn Fadlan", *Rocznik*

orientalistyczny 49 (1995): 165–9; F. Sezgin (compiler), *Islamic Geography 169: Studies on the Travel Accounts of Ibn Fadlan and Abū Dulaf (first half 10th century)* (Frankfurt am Main, 1994); James Montgomary, "Ibn Fadlan and the Rusiyyah", *Journal of Arabic and Islamic Studies* 3 (2000): 1–25.

71 Abū Dulaf, *al-Risāla al-ʿula*, in Fami Saʿd and Ṭalal Majdhūb (eds.), *Taḥqīq al-makhṭūṭāt bayna al-naẓariyya wal-taṭbīq* (Beirut, 1993); ed. Marzin Saʿid Marzin Asiri (Mecca, 1995); A. von Rohr-Sauer, *Abū Dulaf Bericht uber seine Reise nach Turkestan* (Bonn, 1933). *al-Risāla al-thāniya* is devoted solely to travels within Iran: see V. Minorsky, *Abū Dulaf Misʿar ibn Muhalhil's Travels in Iran (circa A.D. 950)* (Cairo, 1955); Yāqūt al-Ḥamāwī al-Rūmī, *Muʿjam al-Buldān*, s.v. *sin*. See also our translation in the second part.

72 Abū Ḥamid Muḥammad al-Gharnaṭī, *Tuḥfat al-albāb wa-nukhbat al-aʿjāb*, ed. G. Ferrand, *Journal Asiatique* 207 (1925): 1–148, 193–304; ed. Alī Umar (Port Said, 2003); Vladimir Minorsky (ed. and trans.), *Marvazi on China, the Turks and India* (London, 1942).

73 G. Browne (trans.), *An Abridged Translation of the History of Tabaristan* (London, 1905), 16–17.

74 Anonymous (fl. *c.* 982), *Ḥudūd al-ʿālam min al-mashriq ilā al-maghrib* (Tehran, 1352/1933); V. Minorsky (trans.), *The Regions of the World: A Persian geography, 372AH/982AD* (London, 1970), ch. 32; Guy Le Strange, *The Lands of The Eastern Caliphate* (London, 1905), 172–3; H.L. Rabino, *Mazandaran and Astarabad* (London, 1928), 1.

75 About the people of Daylam, see Minorsky (trans.), *Ḥudūd al-ʿālam*, 133, 136–37, 384–85, 387–91; *idem*, "La Domination des Dailamites" (Paris, 1932) (reprinted in his *Iranica* (Tehran, 1964), 12–30).

76 Abū Isḥāq Ibrāhīm b. Hilāl al-Ṣābī, *al-Muntazaʿ min kitāb al-tāji li-abi isḥāq al-ṣābi*, ed. al-Zabaydī (Baghdad, 1977), 26, 35 (a description of the history of Daylam based upon a local informant named Aḥmad b. ʿAlī al-ʿAmarī al-ʿAlawī al-Ṭabarī).

77 John J. Donohue, *The Buwayhid Dynasty in Iraq 334 /945 to 403/1012: Shaping Institutions for the Future* (Leiden, 2003).

78 I.M. Lapidus, *A History of Islamic Societies* (Cambridge, 1988), 139, 147.

79 Bahram was the fifth Sasanian king (r. 421-439). Arab sources report that he was brought up at the court of the Arab king al-Mundhir and that he conducted triumphant campaigns in Inner Asia. See al-Ṣābī, *Kitāb al-Tāji fī akhbār al-dawla al-daylamiyya*, in W. Madelung (ed.), *Arabic Texts Concerning the History of the Zaydi Imams of Tabaristan, Daylaman and Gilan* (Beirut, 1987), 12–13; *idem*, "Abū Isḥaq al-Ṣābi on the ʿAlīds of Tabaristan and Gilan", *JNES* 26 (1967): 21, 24.

80 W. Madelung, "The Assumption of the Title Shahanshah by the Buyids and the Reign of the Daylam", *JNES* 28 (1969): 84–108, 168–83; Lutz Richter-Bernburg, "Amir-Malik-Shahanshah: ʿAḍud ad-Daula's Titulature Re-examined", *Iran* 18 (1980): 83–102.

81 Yusuf Ibish, *The Political Doctrine of Baqillani* (Beirut, 1966), 97–8.

82 For a coin minted by ʿAḍud al-Dawla, see S. Lane-Poole, *Catalogue of Oriental Coins in the British Museum* (London, 1875–1890), 2: 207 (Shiraz 338/949); *idem*, *Additions to the Oriental Collection* (London, 1889), 262.

83 Even later generations voiced their admiration for his enterprise: V. Minorsky, "A Civil and Military Review in Fars in 881/1476", *BSOAS* 1 (1939): 152.

84 A site at the end of the Pulvar Valley, five kilometres south to the large mound of the Islamic city of Iṣṭakhr and about sixty from Shiraz.

85 J.J. Donohue, "Three Buwayhid Inscriptions", *Arabica* 20 (1973): 75.

86 *Dīwān*, ed. Dieterici (Berlin, 1861), 762, 769, 792–4; R. Blachere, *Abū Ṭayyib al-Motanabbi* (Paris, 1935), 244–8; Abū Manṣūr ʿAbd Allah b. Muḥammad al-Thaʿālibī al-Naysabūrī (d. 1038), *Yatīmat al-dahr fī [maḥāsin] shuʿarāʾ ahl al-ʿaṣr* (Cairo, 1885), 3: 230, 233.

87 H. Busse, *Chalif und Grosskoning* (Beirut, 1959), 159–62.

88 Cl. Cahen, "Une correpodance Buyid inedite", *Studi Orientalistici in onore di G. Levi Della Vida* (Rome, 1956), 1, 83–97 (dispatches by ʿAbd al-ʿAziz b. Yūsuf); J. Ch. Burgel, *Die Hofcorespondenz Adud al-Dawlas* (Wiesbaden, 1965), 87–9; Ismaʿil Ghalib, *Meskukāt qādima islāmiyya kataluğu* (Istanbul, 1894), 352.

89 Nabih Abbott, "Two Buyid Coins in the Oriental Institute", *American Journal of Semitic Languages and Literatures* 56 (1930): 350–1 (a silver dirham coin from Palestine).

90 G. Wiet, *L'Expedition Persane* (Cairo, 1931), 11, 13.

91 J. Sadan (ed.), *Nouvelle source sur l'epoque Buyide* (Tel Aviv, 1980), 66–7.

92 Klaus S. Hachmeier, "Private Letters, Official Correspondence: Buyid *inshāʾ* as a Historical Source", *Journal of Islamic Studies* 13 (2002): 130–1.

93 Wiet, *L'Expedition Persane*, 31

94 Cahen, "Une correspondance Buyide inedite", 1: 93–5.

95 M.S. Khan, "Miskawaih and Arabic Historiography", *JAOS* 89 (1969): 718, 730; Mathieu Tillier, "L'Exemplarité chez al-Tanùkhì : les cadis dans le *Nishwàr al-Muhadara", Arabica* 44 (2007): 6–7.

96 B. Dodge (trans.), *The Fihrist of Ibn al-Nadīm: A Tenth-century Survey of Muslim Culture* (New York, 1970), 1: 296.

97 Ibrahim Kafesoğlu, "A Propos du nom Türkmen", *Oriens* 11 (1958): 146–50.

98 Tourkhan Gandjeï, "Turkish in Pre-Mongol Persian Poetry", *BSOAS* 49 (1986): 67–75.

99 P.B. Golden, "The Migrations of the Oguz", *Archivum Ottomanicum* 4 (1972): 47; *idem*, *An Introduction to the History of the Turkic People: Ethnogenesis and State-Formation in Medieval and Early Modern Eurasia and the Middle East* (Wiesbaden, 1992), 145, 206.

100 Abū al-Faḍl Muḥammad al-Bayhaqī, *Taʾrīkh-i Masʿūdī* ed. Saʿid Nafisi (Tehran, 1326 shamsi), 2: 653.

101 On this Central Asian dynasty, see Luke Treadwell, "The Account of the Samanid Dynasty in Ibn Ẓāfir al-Azdī's *Akhbār al-duwal al-munqaṭiʿa*", *Iran* 43 (2005): 135–71.

102 Clifford Edmund Bosworth, *The Ghaznavids: Their Empire in Afghanistan and Eastern Iran 994–1040* (Edinburgh, 1963); *idem*, *The Later Ghaznavids: Splendour and Decay 1040–1186* (Edinburgh, 1977).

103 Bull [Taurus] or camel stallion (*faḥl[u] al-ibil*). al-Kāshgharī, *Dīwān lughāt al-turk*, 1: 352 (l. 4), 429 (13), 3: 167 (ll. 8–9); Dankoff (trans.), *Compendium of the Turkic Dialects*, 1: 319, 383, 2: 265; Gerald Clauson, *An Etymological Dictionary of Pre-Thirteenth-Century Turkish* (Oxford, 1972), 312/a; Clifford Edmund Bosworth, "Notes on Some Turkish Names in Abū 'l-Faḍl Bayhaqi's Tārīkh-i Masʿūdi", *Oriens* 36 (2001): 305.

104 M. Hamada, "Le Mausolee et le cults de Satuq Bughara Khan", *Journal of the History of Sufism* 3 (2001) 63–87. It seems that the story of his conversion to the new religion under the influence of a dream is a topos.

105 O. Pritsak, "Von den Karluk zu den Karachaniden", *Zeitschrift der Deutschen Morgenländischen Gesellschaft* 101 (1951): 270–300; *idem*, "Die Karachaniden", *Der Islam* 31 (1953): 17–63.

106 Klaus Lech, *Das Mongolische Weltreich: al-Umari's Darstellung der mongolischen Reiche in seinem Werk* (Wiesbaden, 1968), 39–40 (Arabic).

107 A panegyric chronology that was composed for the Sultan Abū al-Fatḥ Kay Khusrau says: "The ancestor of the Saljūqs was an army commander (*sū-bāshi*) who commanded innumerable soldiers and accomplished all his desires. He had immense property, horses and disciplined retinues. They were mighty already in the land of the Turks. This was so due to the multitude of clans that they governed and the magnificent meadows [that they controlled]." Muḥammad b. ʿAlī b. Sulaymān al-Ravandī (fl. *c.* 1204–5),

Rāḥat al-ṣudūr wa-ayāt al-surūr, ed. Muhammad Iqbal (London, 1921), 87 (ll. 14–17, 22–23).

108 This is clearly reflected in later sources that brought together narratives from earlier accounts. The well-known historian Rashīd al-Dīn describes the leading elite (*khaṣṣa*) and says: "It should be understood that the chosen (*gāzidagān*) among the created beings (*khalā'iq*) are angels, prophets and rulers (*pādishāhān*). The kings (*mulūk*) are the shepherds (*rā'i*) of the flock and the protectors (*ḥamī*) of the creatures from all kinds of calamity and dread. It is the nature of the kings to have knowledge (*'aql*) and ability to govern (*siyāsa*)." Rashīd al-Dīn Faḍl Allah, *Jāmi' al-tavārīkh,* ed. Ateş (Istanbul, 1960) 2/5 [*selcuklular tarihi*]: 2–3; Ẓahīr al-Dīn Nīshāpūri, *The History of the Seljuq Turks: From the Jami al-Tawārīkh – an Ilkhanid Adaption of the Saljūq Nama*, trans. K.A. Luther, ed. C. Edmund Bosworth (London, 2001), 25.

109 Claude Cahen, "Le Malik-Nameh et l'histoire des origines Seljukides", *Oriens* 2 (1949): 42 (Dukak); Donald S. Richards (trans.), *The Annals of the Saljūq Turks: Selections from al-Kāmil fil-Ta'rīkh of 'Izz al-Dīn Ibn al-Athīr* (London, 2002), 31; Omid Safi, *The Politics of Knowledge in Pre-modern Islam: Negotiating Ideology and Religious Inquiry* (Chapel Hill, 2006), 13–15 and see below.

110 R. Dankoff, "Kashghari on the Tribal and Kingship Organization of the Turks", *Archivum Ottomanicum* 4 (1972), 26, 27 (and note 9).

111 al-Kāshgharī explains the name as a courageous falcon (*bazi, ṣaqr*) and adds "from this comes the man's name Chagri B.K (*beg*)". *Dīwān lughāt al-turk*, 1: 347 (ll. 2–3), 352 (11); Dankoff (trans.), *Compendium*, 1: 314, 319.

112 Abū al-Faḍl al-Bayhaqī reports that the vizier Abū Naṣr Mishkān told him in Rajab 426/May 1035: "O Abū al-Faḍl! Khurasān is lost. Go immediately to the chief master and inform him of this." *Ta'rīkh-i mas'ūdī*, 1: 571–2.

113 On the thesis of climate change versus the interpretation that emphasises political conflicts, see Richard W. Bulliet, *Cotton, Climate, and Camels in Early Islamic Iran: A Moment in World History* (New York, 2009).

114 Abū al-Faḍl al-Bayhaqī, *Ta'rīkh-i mas'ūdī*, 1: 572, 612.

115 Niẓām al-Mulk, *Siyar al-mulūk (Siyāsat nāmah)* (Tehran, 1340 sh.), 130; H. Darke (trans.), *The Book of Government or Rules for Kings* (New Haven, 1960), 105.

116 Rashīd al-Dīn, *Jāmi' al-tavārīkh* [*cami al-tavarih*] 2/5: 19; Ẓahīr al-Dīn al-Nīshāpūrī, *The History of the Seljuq Turks*, 40.

117 Omid Safi, *The Politics of Knowledge*, 5.

118 Ibn al-Jawzī, who wrote his chronicle a century after the events, quotes a communiqué (*kitāb*) sent by Ṭughril Beg to the chief judge (*qāḍī al-quḍāt*), which contains his ideology and legitimacy. The letter opens with the lines: "From the exalted King of Kings (*shāh-an shāh*), the King of the East and the West, the Reviver (*muḥyyi*) of Islam, the Lieutenant of the Muslims' Leader (*khalīfat al-imām*), the supporting Right Hand (*yamīn*) of God's viceroy on Earth (*khalīfat allah*) and the Commander of the Faithful (*amīr al-mu'minīn*)." *al-Muntaẓam, fī tawārīkh al-mulūk wal-'umam* (Haydar-Abad, AH1359), ed. S. Zakkar (Beirut, 1995); ed. M. Ata (Beirut, 1992), 9: 382 (year 454/1062).

119 George Makdisi, "The Marriage of Tughril Beg", *IJMES* 1 (1970), 259–75 (reprinted in his *History and Politics in Eleventh Century Baghdad* (London, 1990), art. 9).

120 On this tradition, see Goldziher, *Muslim Studies*, 2: 32–3; Ella Landau-Tasseron, "The 'Cyclical Reform': A Study of the Mujaddid Tradition", *Studia Islamica* 70 (1989): 107–8.

121 al-Tha'ālibī al-Naysabūrī, *Kitāb tatimmat al-yatīma*, 1: 107–12 (the author quotes data from Ibn Ḥassūl); Abū al-Ḥasan 'Alī b. al-Ḥasan b. 'Alī al-Bakharzī (d. 1075–76), *Dumyat al-qaṣr wa-'uṣrat ahl al-'aṣr*, ed. A. al-Fataḥ M. al-Ḥilu (Cairo, 1968), 1: 413–17 (the writer met with Ibn Ḥassūl in Rayy during 1052); Ṣalāḥ al-Dīn Khalīl al-Ṣafadi, *al-Wāfī bil-wafāyāt*, ed. S. Dedering (Damascus, 1959), 4: 132–5 (no. 1642); Muḥammad b. Shākir al-Kutubī (d. 1363), *Fawāt al-Wafāyāt*, ed. I. 'Abbās

(Beirut, 1974), 3: 430–2; Abū al-Ḥasan Jamāl al-Dīn ʿAlī b. Yūsuf al-Qifṭī, *al-Muḥammadūn min al-shuʿrāʾ wa-ashʿaruhum*, ed. H. Maʿammari (Riyad, 1970), 367; Muḥsin al-Amīn, *Aʿyan al-shiʿa* (Beirut, 1986), 9: 444–7.

122 Ẓahīr al-Dīn Nīshāpūrī, *The History of the Seljuq Turks*, 43.

123 Muḥammad Nazim, *The Life and Times of Sultan Mahmud of Ghazna*, 2nd edn (New Delhi, 1971), 81–5; G.C. Miles, *The numismatic history of Rayy* (New York, 1938), 187.

124 Richards (trans.), *The Annals of the Saljūq Turks*, 50.

125 E.G. Browne, "Account of a Rare, if not Unique, Manuscript History of the Seljuqs", *JRAS* (1902), 593.

126 Claude Cahen, "Le Malik-Nameh et l'histoire des origines Seljukides", *Oriens* 2 (1949): 31–65.

127 Hayden White, "The Value of Narrativity in the Representation of Reality", in W.J.T. Mitchell (ed.), *On Narrative* (Chicago, 1981), 1–23.

128 Abū Manṣūr Muḥammad al-Kundurī served Ṭughril Beg for a decade (1054–63). See Shams al-Dīn Ibn Khallikān (1211–1282), *Biographical Dictionary,* trans. Mac Guckin de Slane (London, 1871), 3: 290–95; Omid Safi, *The Politics of Knowledge in Pre-modern Islam,* 52; for his religious policy, see Richard W. Bulliet, "The Shaikh Al-Islam and the Evolution of Islamic Society", *Studia Islamica* 35 (1972): 62.

129 This book, being a narrative of the Daylamites (*fī athār al-dawla al-daylamiyya* or *fī akhbār al-daylam*), contains long phrases in praise of ʿAḍud al-Dawla. The book, known also as *Tawārīkh āl buya* (*buwayh*) was not unfamiliar during the Middle Ages, yet only fragments of this work have reached us. M.S. Khan, "A Manuscript of an Epitome of al-Ṣābi's *Kitāb al-Tajī*", *Arabica* 12 (1965), 27–44; *idem*, "The Contents of the *Kitāb al-Taji*. A Manuscript of Abū Isḥāq Ibrāhām al-Ṣābī", *Islamic Studies* 8 (1968): 247–52; *idem*, "Studies in the *Kitāb al-Tāji* of al-Ṣābī", *Arabica* 17 (1970), 151–60; W. Madelung, "Abū Isḥāq al-Ṣābī on the ʿAlīds of Tabaristan and Gilan", *JNES* 26 (1967), 17–57; *idem*, "Notes on al-Ṣābī's *Kitāb al-Tajī*", *Islamic Studies* 9 (1970): 81–8. Excerpts (*muntazaʿāt*) from the first part of *Kitāb al-tāji fī akhbār al-dawla al-daylamiyya* were published by M.H. Al-Zabaydi (Baghdad, 1977) and by W. Madelung (ed.), *Arabic Texts Concerning the History of the Zaydi Imams of Tabaristan,Daylaman and Gilan* (Beirut, 1987). This book was well known among Persian and Arab writers of later periods. E.G. Browne (trans.), *An Abridged Translation of the History of Tabaristan* (London, 1905), 90.

130 But see Niẓām al-Mulk, *Siyāsat Nāmah*, 96. Cf. also Ibn Isfandiyar, Baha al-Dīn Muḥammad, *Taʾrīkh-i Ṯabaristan* ed. ʿAbbās Iqbāl (Tehran 1350 *shamsi*), 141 (Browne (trans.), *An Abridged Translation of the History of Tabaristan*, 91).

131 S. Shaked, "From Iran to Islam: On Some Symbols of Royalty", *JSAI* 7 (1986): 75–6 (reprinted in *idem*, *From Zoroastrian Iran to Islam: Studies in Religious History and Intercultural Contacts* (Aldershot, 1995) art. 7); R. Frey, *The Golden Age of Persia* (London, 1975), 20, 111; R. Hillenbrand, *Islamic Architecture* (New York, 1994), 390, 392, 393.

132 Jan Rypka, *A History of Iranian Literature* (Dordrecht, 1968), 67, 126–35.

133 For the image of Alexander and Ardashir in medieval sources, cf. the preface by Abū al-Faḍl al-Bayhaqī: "I say, that the most excellent past kings are a group who were the greatest [kings]. And from that group two individuals are well known: one, Alexander the Greek; and the other, Ardashir the Persian. Since our [Ghaznavid] masters and emperors (*padshāh*) have surpassed these two in all things, one must of necessity know that our kings have been the greatest on earth, because Alexander was a man whose princely flame took hold forcefully and flared up for only a very little while and then was reduced to ashes." Abū al-Faḍl Muḥammad ibn Ḥusayn [Kātib/Dabīr] Bayhaqī (996–1077), *Taʾrīkh-i masʿūdī, maʿrūf bih Tārīkh-i bayhaqi*, ed. M.G. Ya Haqqi (Mashhad, 1383/2002), 118; Marilyn Robinson Waldman, *Toward a Theory of*

Historical Narrative: A Case Study in Perso-Islamicate Historiography (Columbus, OH, 1980), 81, 153.

134 Qur'ān, The Prophets, 21: 96.

135 S.M. Stern, "The Arabic Translation of the Pseudo-Aristotelian Treatise *De Mundo*", *Le Museon* 77 (1964): 193; Joel L. Kraemer, "Humanism in the Renaissance of Islam: A Preliminary Study", *JAOS* 104 (1984): 153, 155.

136 Julie Scott Meisami, *Persian Historiography to the End of the Twelfth Century* (Edinburgh, 1999), 210.

137 Nasser O. Rabbat, "The Ideological Significance of the Dar al-'Adl in the Medieval Islamic Orient", *IJMES* 27 (1995): 6.

138 Ibn al-'Adim, Kamāl al-Dīn 'Umar ibn Aḥmad Ibn Abī Jarāda (588–660/1192–1262), *Zubdat al-ḥalab min ta'rīkh ḥalab*, ed. S. Dahhan (Damascus, 1954), 2: 20 (ll. 1–6); *idem*, *Bughyat al-ṭalab fī ta'rīkh ḥalab* (facsimile edition; Frankfurt, 1989), 4: 543; Taqī al-Dīn Aḥmad b. 'Alī al-Maqrīzī (1364–1441), *Kitāb al-Muqaffā al-kabīr*, ed. M. al-Ya'lawi (Beirut, 1411/1991), 2: 220.

139 David Ayalon, "The Mamluks of the Seljuks: Islam's Military Might at the Crossroads", *JRAS* (1996): 314–15, 317–18; Carole Hillenbrand, *Turkish Myth and Muslim Symbol: The Battle of Manzikert* (Edinburgh, 2007).

140 Niẓām al-Mulk, *Siyāsat Nāmah*, 13; Darke (trans.), *The Book of Government*, 9, 11, 68. This position resembles in a great measure the view expressed by Ibn Ḥassūl and hence it may reflect official Saljūq policy. On this book see Neguin Yavari, "Mirrors for Princes or a Hall of Mirrors? Niẓām al-Mulk's Siyar al-mulūk Reconsidered", *al-Masāq* 20/1 (2008): 47–69.

141 Abū Ḥāmid al-Ghazālī, *Naṣīḥat al-mulūk*, ed. R. Bagley (Oxford, 1964), 45, 46, 71, 106.

142 Ibn Bābā's book on the Saljūqs did not survive, nor did its abridgement by Ḥuṣām al-Dīn Abū al-Wafā Muḥammad b. Muḥammad b. 'Umar [Qāsim] al-Akhsikatī (Akhsikathi) Dhū al-Manāqib (d. 1247); Ḥajji Khalīfa (Kâtip Çelebi, 1609–1657), *Kashf al-ẓunūn* (Istanbul, 1940), 1: 281; Viqar Ahmed Hamdani, "Some Rare Manuscripts in Istanbul", *JRAS* (1938): 561–4; C.E. Bosworth, "Ibn Bābā", *EI*[2], 4: 696 a–b; *idem*, "Ebn Bābā Kāšāni (Qāšāni), Abū'l-'Abbās", *Encyclopedia Iranica*.

143 Abū al-'Abbās Aḥmad b. 'Alī Ibn Bābah al-Qāshānī, *Rās māl al-nadīm fī tawārīkh a'ayān ahl al-islām*, ed. S. Zakkār (Beirut, 1418/1997), 264, 266.

144 Meisami, *Persian Historiography to the End of the Twelfth Century*, 188–209.

145 C.E. Bosworth, "Abū Manṣūr Farāmarz", *Encyclopedia Iranica*.

146 Clifford Edmund Bosworth (trans.), *The History of the Seljuq State: A Translation with Commentary of the 'Akhbār al-dawla al-saljūqiyya'* (London, 2011), 20.

147 Mahmud Omid Salar (ed.), *Mujmal al-tavārīkh wal-qiṣaṣ* (facsimile of the Berlin manuscript; Teheran, 2001), fol. 145 (ed. M.T. Bahar (repr.Teheran, 2003), 406–7).

148 Ẓahīr al-Dīn al-Nīshāpūri, *The History of the Seljuq Turks*, 29.

149 Bosworth, *The Ghaznavids*, 268.

150 Richards (trans.), *The Annals of the Saljūq Turks*.

151 Claude Cahen, "The Historiography of the Seljuqid Period", in B. Lewis and P.M. Holt (eds.), *Historians of the Middle East* (London, 1962), 59–78.

152 This orthography is clearly readable in the Syriac text. Mark Dickens, *Medieval Syriac Historians' Perception of the Turks* (MPhil dissertation in Aramaic and Syriac Studies, University of Cambridge, Faculty of Oriental Studies 2004), 15–16.

153 See his own words in the chapter published by Cl. Cahen, "*Akhbār al-Ayyūbiyyin*", *BEO* 15 (1967), 127 (ll. 1–3).

154 Suleymaniye library in Istanbul: Laleli ms. 2002 *kitab İbnü'l Amid'e ait Tarih-i Muhtasari't Taberi*. The Laleli catalogue dates the composition of the work to 1260. It was copied in 1355. The catalogue number 1002 given by Cahen seems to be wrong.

155 Due to a bilingual edition, this history has been known in Europe since the seventeenth century: el-Makān Jirjis ibn al-'Amīd, *Historia saracenica, qua res gestae Muslimorum*

inde a Muhammede arabe reddita opera & stidio Thomæ Erpenii (Erpenius) (Leiden, 1625).The Oxford manuscripts were recently re-edited by A.B. Hassan (Cairo, 2010).

156 *al-Ta'rīkh al-ṣāliḥi* (written in 1248), Istanbul, Suleymaniye Library, Fatih ms. 4224.

157 Claude Cahen, "Al-Makīn ibn al-'Amīd et l'historiographie musulmane: un cas d'interpénétration confessionnelle", in J.M. Barral (ed.), *Orientalia Hispanica sive studia F. M. Pareja octogenario dedicata* (Leiden, 1974), 158–61.

158 Meisami, *Persian Historiography to the End of the Twelfth Century*, 211–13; Parvaneh Pourshariati, "Local Historiography in Early Medieval Iran and the Tārīkh-i Bayhaq", *Iranian Studies* 33 (2000): 145.

159 David Ayalon, "The European-Asiatic Steppe: A Major Reservoir of Power for the Islamic World", in *The Proceedings of the 25th Congress of Orientalists, Moscow 1960*, (Moscow, 1963), 2: 47–52 (reprinted in his *The Mamluk Military Society: Collected Studies*, (London 1979), art. VIII); *idem*, "Mamlukiyat", *JSAI* 2 (1980): 335–7.

160 Baki Tezcan, "Hanafism and the Turks in al-Ṭarasūsī's Gift for the Turks (1352)", *Mamluk Studies Review* 15 (2011): 67–86.

161 Gunhild Graf, *Die Epitome der Universalchronik Ibn ad-Dawadaris im Verhaltnis zur Langfassung: eine quellenkritische Studie zur Geschichte der aegyptischen Mamluken* (Berlin, 1990).

162 Li Guo, "Mamluk Historiographic Studies: The State of the Art", *MSR* 1 (1997): 23.

163 Robert Irwin, "Mamluk History and historians", in Roger Allen and D. S. Richards (eds.), *The Cambridge History of Arabic Literature: Arabic Literature in the Post-Classical Period* (Cambridge, 2008), 164.

164 Faruk Suemer, Ahmet E. Uysal and Warren S. Walker (trans.), *The Book of Dede Korkut: A Turkish Epic* (Austin, 1972), xii.

165 Otfried Weintritt, *Formen spatmittelalterlicher islamischer Geschichtsdarstellung: Untersuchungen zu an-Nuwairi al-Iskandaranis Kitab al-Ilmam und verwandten zeitgenossischen Texten* (Beirut, 1992), 185–6; Anne F. Broadbridge, "Academic Rivalry and the Patronage System in Fifteenth-Century Egypt: al-'Ayni, al-Maqrizi and Ibn Hajar al-'Asqlani", *MSR* 3 (1999): 95.

166 Irwin, "Mamluk History and Historians", 168.

167 Altan Çetin, "Oghuz Turks in the Account of Mamluk Historians", *Journal of Islamic Studies* 20/3 (2009): 376–82.

168 John Masson Smith, Jr. "Turanian Nomadism and Iranian Politics", *Iranian Studies* 11 (1978): 64–7.

169 Charles Melville, "History: From the Saljūqs to the Aq-Qoyunlu (ca. 1000–1500 C.E.)", *Iranian Studies* 31 (1998): 473–4.

Part II

Translated sources

1 THE EURASIAN STEPPES BEFORE THE SALJŪQS

a) Ibn Khurradādhbih: The Central Asian frontiers (ninth century)[1]

Little is known about Abū al-Qāsim ʿUbayd Allāh ibn Khurradādhbih (or Khurdādhbih; the Excellent Gift of the Sun; c. 820–911). Based upon the limited data it is possible to depict him as the offspring of an Iranian family that converted to Islam during the early ʿAbbāsid decades. He served the administration of this dynasty. In this capacity he composed a manual of roads, provinces and taxes. The first version of it was presented to the caliph al-Wāthiq (r. 842–847).[2] The account of the Turkic peoples translated here is extracted from this work. Moving from Iran to the capital Samarra, he became a *nadīm* (boon companion or confidant) of the caliph al-Muʿtamid (r. 870–892). Among the titles that are ascribed to him is a history of the pre-Islamic Iranian dynasties, including the story of Alexander the Great.[3] The Turkic peoples are also mentioned in that book.

<30> The journey from Upper Nūshajān to the [capital] city of the king (*khāqān*)[4] Toghuz-Ghuz[5] lasts three months. The road traverses large and productive villages. The population along this road is *atrāk*, among them are Zoroastrians, who worship fire, and Manichaeans. The king resides in a great walled city with twelve iron gates. The inhabitants are Manichaeans. On the left (i.e. northern) side of the road live the Kaymāk (read Kīmāk i.e. Kīmäk), and ahead eastward, at a distance of about three hundred parasangs,[6] lies China. On his palace's roof the king of the Toghuz-Ghuz has a golden pavilion. It is big and can admit one hundred persons. It is visible from as far as five parasangs.[7] The king of the Kīmāk dwells in camp tents and follows the grass in his wandering. It takes eighty-one days across the steppes to travel the distance between his headquarters and Ṭarāz.[8]

<31> The land populated by the Toghuz-Ghuz is the larger Turkic territory. It is on the southern side of the [Syr Darya] River. They are bordered by China and Tibet [and the] Kharlukh (Qarluq),[9] Kīmäk, Ghuzz, Jifar (Jiqar or Jikil/Cigil), Bajānāk (Pečenegs), Turkash (Türgesh), Arkash, Khifshākh (Qipčaq) and Khirkhīz (Ḳirgiz) Turkic tribes. Musk is [abundant] in that land. The Kharlukh

and the Khalaj reside on the southern bank of the Syr Darya (Jaxartes) River. There are two frontier posts (*maslaḥa*) in the town of Fārāb (Otrār) [in Transoxania], one is occupied by Muslims, the second by Kharlukh (Qarluq). All together there are sixteen cities in the lands of the Turks.[10]

<261> Frontier regions (*thughūr*) encircle the Abode of Islam.[11] One of the largest is the frontier area of the Turks. The Turks inhabit a prairie north of the province of Jurjān (Gurgan)[12] and from there they raid the Abode of Islam. To prevent the Turks' incursion the people of Jurjān used baked bricks to build a wall, yet the Turks captured the wall and conquered Jurjan. They were headed by a king named Ṣūl (Chul/Shul).[13] After an unspecified time the Muslims conquered Jurjān.

Nūshajān is the principal place of the Turks in the frontier area. It is located beyond Samarqand, about sixty parasangs eastwards, in the direction of Shahsh and Farghāna. This is the first line of fortification of the Kharlukhiyya (Qarluq; Karluklar in modern Turkish; Khallukh in Persian),[14] which stretches till the border with the Kīmāk (Kīmäk). From that frontier's crossing point to the city of the Ṭughuz-Ghuzz it takes forty-five days to journey across the prairie. The first twenty days are in grassland, a region with springs of water; the second twenty-five days are across a zone of big villages. The majority of the population of these villages is Zoroastrians. Some among them are heretics who adhere to Mazdakism. Near the city of the Ṭughuz-Ghuzz is a lake that is surrounded by villages and uninterrupted construction. The city has twelve iron gates and is protected by the entire Turkic people. Most of them adhere to Mazdakism.

Upper Nūshajān is forty stages from the city of Shāsh. Convoys will cover the distance of forty stages [in forty days], while a fast traveller will do it in thirty days. Four big towns and five small ones compose Upper Nūshajān. The fighting army of Upper Nushajan encamps in one of these towns, which is built on the banks of a lake. They number twenty thousand fighters. They all are registered in a single inventory (*dīwān*). Among the Turks there are no more valiant fighters than that battalion. Ten of them will overpower one hundred Kharlukhiyya. The town of the Ṭughuz-Ghuzz is built on the shores of a lake whose other sides are enclosed by mountains.[15]

The land of the Kīmāk (Kimäk)[16] is the distance of sixty-five parasangs from Lower Nūshajān,[17] which is beyond Samarkand as we have already said, northward towards the left.[18] It takes eighty days to travel from Ṭarāz (Talas) to the Kimäk.[19] The road crosses vast deserts and grasslands, rich in water and grass. The Muslims are prevented from attacking the Turks, since the Prophet has said: "Leave the Turks alone as long as they do not trouble you." Yet we mentioned them here and described their manners because we undertook to provide a comprehensive account of the lands that encircle the Abode of Islam and of the people that oppose the Muslims.

b) Ibn Faqīh al-Hamadhānī: On the Turks and their lands[20]

The biography of this Iranian author of a geography work is little known. Sprenger assumes that Ibn Faqīh composed his book around the year AD 903.[21] From Arab sources we can deduce that his work circulated in the ʿAbbāsid domains (in the second half of the tenth century). There are two recensions of this work: an abridged one, which reports very poorly on the Eurasian steppes, and a fuller manuscript that was studied by Validi Togan in Mashhad.[22] This text narrates the story of the Turkic people.

<633> Khudhayfa transmitted an apocalyptic tradition: "The Turks will capture Kufa, the Khazar will conquer Northern Iraq (*al-Jazīra*) and the Byzantines will occupy Syria."

The Prophet Muḥammad said: "The Turks will expel the people of Iraq from their land."[23]

It is reported that the caliph ʿUmar wrote to his provincial governors (*ʿummāl*) ordering them: "If you meet a Turk slay him, because they will revolt after the year AH200 (10 August 815–30 July 816). When they revolt they will be more eager to grab your possessions than to protect what they hold."

The Prophet Muḥammad said: "The Turks will seize the first opportunity to snatch from my community whatever they grasp."

It was transmitted from ʿAbd Allāh Ibn ʿAbbās that the Prophet had said: "The caliphate will be in the hands of my progeny until their glory is seized by the red faces whose broad, round faces resemble iron shields."

A *ḥadīth* transmitted by Abū Hurayra says: "The Hour of the Resurrection (*sāʿa*) shall not come until a nation with wide faces, narrow small eyes and flat noses ties their horses along the banks of the Tigris River."

[The caliph] Muʿāwiya said: "Let sleeping dogs lie! Avoid them as long as they do not trouble you; namely do not disturb the Turks and the Ethiopians."

A tradition traced back to the Prophet (*ḥadīth marfūʿ*) says: "The Prophet Muḥammad said: Leave the Turks as long they leave you in peace."[24]

It is said that in the land of the Turks a ewe does not deliver less than four, and sometimes even as many as five or six, lambs, like a she-dog. The delivery of two or three [lambs] is rare and these offspring are truly large in size. These sheep are indeed huge and have an immense tail, which drags on the ground.

<634> The land of the Turks incorporates several regions.[25] The largest territory is that of the Ṭoghuz-Oghuz.[26] They are surrounded by China and Tibet and by other Turkic groups: the Kharlukh (Khallukh; Qarluq), Kīmäk (Yemek),[27] Ghuzz, Cigil,[28] Pečenegs,[29] Turgesh, Arkash, Khifkhākh (Qipčaq)[30] and Khirkhīz (Qirghiz).[31] In this land musk is [plentiful]. This region is on the southern bank of the Syr Darya (Jaxartes) River. In Fārāb there is a frontier castle (*maslaḥa*)[32] where a Muslim unit is stationed.[33] A second fortress is manned by a force of Qarluq soldiers. In the lands of the Turks are sixteen cities.

Experts in Turkic affairs divide them into several races (genus) of people (*ajnās*): 1) The Qarluqs live close to the province of Samarqand. They are the best Turkic people; 2) The Badhkashiya [read: *Badakhshan*] have long beards; 3) The Oghuz and Toghuz; 4) The Kīmāks are the leading tribe. They live deep in their country and are regarded by the other Turkic people as the most courageous; 5) The Bushnayika (Pečenegs?) and the Shariya (Shaghiya or Shārī?).

The Toghuz-Oghuz people are the Arabs of the Turks (*ʿarab al-turk*).[34] They own meadows. Erecting and dismantling [tents], they wander. The Badhkashiya people are sedentary and dwell in houses and villages.

An account narrates a story about the Umayyad caliph Hishām b. ʿAbd al-Malik (r. 724–743) who sent an envoy to the king of the Turks, calling him to embrace Islam. The messenger said: "I approached the king while he was crafting a horse's saddle (or saddling)." When the king saw the foreigner he asked the interpreter: "Who is this man?" At this the interpreter responded: "An ambassador from the king of the Arabs". The king asked: "Is he a young man (*ghulām*)?" "Yes", was the answer. He ordered me to be conducted to a residence where plentiful meat and little bread were served.[35]

[Several days later the king called me and asked: "What is your purpose?" I answered politely: "My master thinks that you are wrong and wishes to give you sound advice. You must convert and accept Islam." The king asked: "What is Islam?" I informed him of its principles, and explained to him the duties and restrictions, rules and rituals. He left me for several days and then he came back.]

Later during that very day the king went riding. He was accompanied by ten commanders, each of whom was holding a flag. He ordered me to join his entourage and I was taken, mounted on the back of a riding animal. <635> The squad claimed a hill [covered with trees], which was in the middle of a swamp. When the sun rose, the king ordered one of his chief commanders to raise a flag and wave it. The officer executed this command. Ten thousand fully armed fighters emerged at this moment [shouting together *jāh jāh*].

[They stopped at the foot of the mountain and their commander mounted the hill and put a crown on the king's head. The king ordered his commanders to raise their flags, and they did so, one after the other. At each call, ten thousand armed riders gathered. This went on till all ten waved their flags. It ended with a gathering of a cavalry force of one hundred thousand riders.] Then the king said to the translator: "Tell the envoy that he should inform his lord that not one of these troops is a cupper, cobbler or tailor. If they convert to Islam how they would earn their food?"

The extreme end of the province of Khorasan (Khurāsān)[36] borders Upper Nūshajān in the province of Shāsh.[37] It takes three months to cover the distance between Upper Nushajan and the capital city of the king (*khākān*) of the Toghuz-Oghuz. The road traverses large villages, fertile fields and markets. The population of this country are Turks, yet among them reside Zoroastrians that worship at fire temples and Manicheans, adherents of Mani who renounced the old Zoroastrian organised religion (*zanādiqa*). The court of the king is in a large and populous city with rich markets. In the city wall there are twelve iron gates.

[The land of the Kīmāk[38] lies to the north of it and China is three hundred parasangs eastward.] Their country is a barren land. The king of the Kīmāk and his people observe the rains, and in search of grass they wander to the region where dew falls.

ʿAlī b. [Sahl] Rabbān [al-Ṭabarī], the secretary of al-Māziyār b. Qārin (d. 841)[39] said: "[The capital city of the Toghuz-Oghuz is strongly fortified.] There is no other city on earth so firmly barricaded. One of the Turks' kings visited a remote location of his kingdom. There was a vast swamp there, whose water was brackish. He ordered the water to be diverted and deep and wide foundations to be excavated. They measured forty cubits (*dhirāʿ*). Using stone and gypsum, he built ramparts. Two wide stockades were erected at the bottom of the deep trench, each of them ten cubits wide. A twenty cubit-wide ditch separated the supporting walls. When this was accomplished and the construction reached the ground level, the ditch was filled (*ṭamma* or *ḍamma*) with sand [and mud]. A mound was thrown up, filled with sand, between the dikes. Then construction went on till the ramparts had risen fifty cubits above the ground. As the mound was advancing, sand was thrown between the dikes. Next, he constructed a city for himself and his retinue in the middle of the ramparts. He built palaces and residences. Around the dike he dug a moat. It was flooded with water. It did not take long, and about a year later the old swamp re-emerged. It could be considered one of the biggest lakes. In this way the strongest city was constructed. The King transported his family and treasury to the place. There is no city in a mountain range or in the plains that is more fortified."

It is well known that no nation can compare to the Turks' skill in piercing walls. They are capable of penetrating cities and fortifications by digging from remote positions. It happened that a Turkic king attempted to capture this city. He advanced toward the place and encamped near the swamp, about one parasang from the city. He ordered his engineers to pierce the fortifications. <636> However, when these engineers reached the swamp that surrounds the place, its water burst through the breach. They persevered in struggling against the current. Time and again the water would flood them; seldom were they able to overcome the water. Eventually they reached solid ground. They believed at that point that they would conquer the city. They turned to the wall and started to pierce it. However, by the time they completed the breach, plenty of sand had fallen on them. They were not able to remove a grain of sand without coming under a heavier flow of silt. Realising this, they concluded that there was no way to capture the city and withdrew, disappointed.

It is reported that in the land of the Turks a ewe would carry several lambs in one womb and occasionally deliver five, six and even seven lambs. Ewes would regularly deliver three or four lambs. Their entire livestock reproduces in a similar manner.[40]

When Turks wish to swear alliance they will produce a deity made of copper and display it. Then, they place in front of the idol a sizeable dish and fill it up with water. In the dish they put a piece of gold and a spoonful of millet and under the dish they arrange a pair of women's trousers. Then they turn to the person taking the oath and say to him: "Swear that if you break the pact, deceive or betray, God will transform you into a woman who would wear these trousers, and He will help someone to subdue you and cut you into small pieces the size of these millet seeds. Your colour will change and become yellow, resembling the colour of the gold." After taking the oath the person would drink the water. No one among them cheats on this oath without being overcome by death or calamity.

In the lands of the Turks there are many excellent sables (*sammūr*) and foxes (*fanak;* fennec) [ideal for making excellent furs].

[Among all the nations the Turks are the most skilled bowmen (*rumat al-ḥadaq*)], no people rival their qualifications as archers.

The Turkic father brings up his child from birth to youth. He takes care of him, supports him and looks after him till the child attains puberty. When the lad reaches maturity, his father provides him with a bow and arrows and sends him away from his house, saying to him: Take care of yourself. From that moment onward, the father treats his son as a complete stranger. This is the Turkic manner of rearing and bringing up their children. [Moreover, among the Turks are groups that sell their sons and daughters into slavery.]

[The young unmarried women do not cover their heads.] The Turkic marriage custom is related to the fact that the young women do not cover their hair. If a man wishes to marry, he looks for a girl till he finds one that attracts his attention, and then he throws a scarf on her head. By doing so she becomes his wife. No one can stop him, neither the girl's father nor her brother.

<637> In the Turkic land there are excellent horns (*khutuw*).[41] They are obtained from the forehead of an animal that they hunt.

Tamīm b. Baḥr al-Muṭṭawwaʿī reports that the country of the Turks is very cold.[42] One can travel in it only during six months of the year.[43] He says that he journeyed to the country of the king (*khāqān*) of Toghuz-Ghuz on relay horses that the *khāqān* sent him. Travelling as hard and as fast as he could, he covered three stages in a day and a night. He journeyed for twenty days in steppes where there were springs and grass but no villages or towns, only the men of the relay service living in tents. He was carrying with him twenty days' provisions, because he knew the affairs of the road to that city and [knew] that the distance was twenty days along the steppes with [only] wells and grass. After that, he travelled for twenty days among villages lying closely together and plentiful cultivated tracts. The people, all of them, or most of them, were Turks. Among them were fire-worshippers professing the Manicheaen religion and adherents of the dualistic heresy (*zanādiqa*).[44]

After all these days he arrived at the king's town. He reports that this is a great town, rich in agriculture and surrounded by rustic counties full of cultivation and villages lying close together. The town has twelve iron gates of a huge size. He narrated that the town is populous and thickly crowded and has markets and various merchandises. Among its population, the dualistic religion of crypto-Manichaeism prevails. He mentioned that he estimated how far from China (*bilād al-ṣīn*) the city is and he worked out that it was a distance of 300 parasangs, adding: "I think it is (even) more than that." He reported that to the south of the king of the Toghuz-Ghuz's town are the lands of the Turks with whom no one else mixes. To the north are the lands of the Kīmäk. Eastward is China.

He mentioned that from a distance of five parasangs before he arrived in the town [of the *khāqān*] he caught sight of a tent belonging to the king. It is made of gold and [stands] on the flat top of king's palace. It can hold nine hundred persons.[45] He records that the *khāqān*, the king of the Toghuz-Ghuz, is related by marriage to the king of China, and that the king of China sends him 500,000 pieces of silk (*farand*) annually.

He also reported that the journey between Upper Nūshajān and al-Shāsh, via Ṭarāz, takes forty days for caravans, but a single rider who journeys on horseback can cross that distance in a month.[46]

He [Tamīm] said that in Upper Nūshajān there are four large towns and four small ones.[47] He estimated the number of warriors in <638> one town there that lies on the shores of a lake, putting them at about 20,000 fully armed horsemen. Among all the Turkic tribes there is no one stronger than they. When they go to war with the Kharlukh (Qarluq), they are a hundred and the Kharlukh a thousand, and [in spite of] this they come out [the winners] in all their wars. He reports that this lake is like a square-shaped pond and around it are high mountains, with all kinds of trees in them.

He [Tamīm][48] said: "And here are traces of an ancient town. I could not find among the Turks anyone knowing its story, who built it, who its people were and when it became ruined." He inspected it up to the site of a river which crosses it. One cannot reach its bottom here. [Tamīm] added: "And I saw in it various sea-creatures which I had not seen before. And I also saw birds, the like of which I had not seen in any of the countries." He [Tamīm] says: "The people of Nūshajān and others from neighbouring towns and villages circumambulate the lake once a year in the spring and they do this as a kind of baptism rite. From the direction of Tibet 150 large and small water-streams enter it, and also from the land of the Toghuz-Ghuz and Kīmāk."

He [Tamīm][49] said that when he travelled to the king of the Toghuz-Ghuz, "he found him encamped in the neighbourhood of his town. He estimated the army that camped around the king's tent, to say nothing of the others, was some 12,000 strong. Besides these there were seventeen chieftains, each having three thousand warriors of his own. Between each two chieftains lay the arsenal tents. The chieftains, together with those arsenals, were encircled by the army. In this circle there is a gap the size of four gates in the direction of the army. All the horses of the king and the troops pasture between the tent of the king and the places occupied by the chieftains, and not one animal escapes outside the camp."

And we asked him about the road from Ṭarāz to the Kīmāk. He mentioned that "from Ṭarāz one travels to two villages called Kawākib (stars),[50] which are flourishing and populous. Their distance from Ṭarāz is seven parasangs, <639> and from this place to the residence of the king of the Kīmāk a hard rider carrying his provisions travels eighty days." These wastelands, steppes, and wildernesses are vast and abound in grass and wells, and in them are the pastures of the Kīmāk peoples. He revealed that he travelled along this road and found the king and his army camping in tents. In the neighbourhood of the encampment were villages and cultivated tracts. The king travels from one place to another, following the grazing. Their animals are numerous and with tiny hoofs. He reckoned those in the king's army and found they were some 20,000 horsemen.[51]

Abū al-Faḍl [ʿAlī b. Muḥammad] al-Wāshjirdī[52] reports that the king of the Toghuz-Ghuz twice raided the kingdom of China. It happened in the days of [Hārūn] al-Rashīd (786–809); however, some argue that it was in the days of [his father] al-Mahdī (775–785). He led an army against China. This foray took place between S.rushna [read: Usrūshana][53] and Samarqand. The governor (*ʿāmil*) of Samarqand fought with the king on several occasions, and some of the battles were heavy. Eventually God granted the ruler (*ṣāḥib*) of Samarqand victory and he defeated the king, killing many of his companions. It is said that the king led an army of about 600,000 Chinese, horsemen and foot soldiers. The Muslims took enormous booty and captured some [of the Chinese] people; their children are those who in Samarqand fabricate good paper and various kinds of arms and implements.[54] These are produced only in Samarqand, of all the towns of Khorasan.

And one of the wonders of the country of the Turks is some pebbles they have, with which they bring down rain, snow, cold, etc., as they wish. The story of these pebbles is well known and widely spread among them and no Turk denies it. And these pebbles are uniquely in the possession of the king of the Toghuz-Ghuz and no other of their kings possesses them.

Abū ʿAbd Allāh al-Ḥusayn b. Ustādhūya told me from Abū Isḥāq Ibrāhīm b. al-Ḥasan → Hishām (d. 819) b. [Muḥammad][55] al-Sāʾib al-Kalbī (d. 763) → Abū Malīḥ → Ibn ʿAbbās as follows: "As long as Sarah was alive Abraham,[56] peace be upon him, did not marry another woman. After her death Abraham married a woman from the original Arabs called Qanṭūrá bint Maqṭūr.[57] She bore him five sons:[58] Midyan,[59] Madayin who is Medan,[60] Nīsān, Ishbaq[61] and Saraj. Abraham, peace be upon him, ordered the progeny of Ishmael, Isaac, Midyan and Nīsān to join him, and expelled Medan, Ishbaq and Saraj. They said to him: 'O father, how is it possible that you permit Ishmael, Isaac, Midyan and Nīsān to stay with you, and to live in security and comfort, while you expel us to faraway lands, <640> to wilderness and loneliness?' Abraham responded: 'So I was ordered. Yet I will provide you with a name, one of Allah's names, and this will provide you with the power to prevail over your enemies and to bring down rain'. He told them this and they started travelling until they settled in a place in Khorasan (Khurāsān) where they multiplied, and using this name, they subjugated all those who opposed them.[62] Their story reached the Khazar, who are descended from the son of Japhet, the son of Noah. They traveled to meet Abraham's progeny, made a pact with them and intermarried with them. Some of the Khazars stayed with them and the remainder returned to their country."[63]

Abū al-ʿAbbās ʿĪsá b. Muḥammad b. ʿĪsá al-Marwazī[64] narrated that in the frontier (*thughūr*) of the province of Khorasan, which lies beyond the Oxus River, and in the other districts that are adjacent to the land of the infidel Turks, amazing stories have been told about the Turks that belonged to the groups of the Ghuzz, Toghuz-Oghuz and Qarluq (Khallukh). He said: "These people consider themselves a powerful entity and a force that can inflict grave harm upon its enemies."

[We were told that] among the Turks some individuals are blessed with the power to trigger the falling of precipitation, while travelling or during other circumstances. Precipitations will fall and from heaven rain, hail, snow or something else will pour down, as they wish. We hesitated to accept these accounts, some believing in the validity of these stories and others rejecting them.

This disagreement prevailed till I met with Dā'ūd b. Manṣūr Ibn Abī ʿAlī al-Bādhghīsī.[65] He was an honest and respectable man who had served as a governor of Khorasan. He did it very efficiently. During his term in office, Dā'ūd b. Manṣūr Ibn Abī ʿAlī al-Bādhghīsī had met privately with Bālqīq b. Ḥayyūwiya, the king of the Ghuzz Turks. During one of their encounters the governor asked the king: "We were told that among the Turks there are individuals who are blessed with the power to generate precipitation. They can cause the pouring down of rain, hail and snow whenever they wish. What do you know about this?"

The king told him: "Indeed, God regards the Turks as the most miserable and lowliest people. As such they would not be able to produce these sorts of miracles. However, the story that you were told is true. There is a reason behind it. Let me tell it to you. One of my forefathers had quarrelled with his father, who at that time was the king of the Turks. He distanced himself from his father and departed from [the court] with servants (*ghilmān*), followers (*mawāli*)[66] and vagabonds who liked loafing. They wandered around and traversed the land, roving in all directions, attacking communities and hunting what they found on their way.

Their journey ended in a land whose population told them that it was impossible to pass beyond that region and that no one can continue travelling, since the way is barred by a mountain. "Why?" [the young prince] asked them. He was told by the inhabitants that the sun rises behind this mountain; it is very near to the earth, so that the sun's rays fall upon everything and burn it. So he asked them: "Do any people or animals inhabit that land?" They replied: "Yes." Since this was contrary to the information that they had told him, he asked them: "How can they live there in the hazardous conditions that you have described?"

They told him that the people that inhabit this land have underground tunnels and caves in the mountains. When the sun rises, they hasten to find shelter <641> and go into these holes and hide there until the sun moves away from them [and then they come out]. And as for the animals, they pick up small stones, of which they have inspired knowledge regarding their qualities. Each animal takes a stone in its mouth and raises its head towards the heavens and a cloud casts a shadow over them and shelters them from the burning sun. The cloud forms a screen between the animals and the sun and this barrier stops the sun's rays."

[Bālqīq] continued his story and added: "My ancestor entered that region and found that the information that had been given to him was indeed accurate, exactly as he had been told. When the sun was rising, all the animals rushed towards these stones and took them in their mouth. Each creature held a stone with his lips. They raised their heads towards the sky and the clouds shadowed them. He [Bālqīq's ancestor] and his companions chased the animals and when their prey became weary, they abandoned the stones. He ordered his men to collect the stones and investigate them. This was done. The stones that the animals dropped from their mouths were gathered and examined. Then he and his men continued to roam this vast land carrying these stones with them and raising them against the sun. The stones caused clouds that sheltered the people from the rays. In this way they escaped the heat of the sun.

"Then they collected as many of these stones as they could and returned to their homeland, using the stones during their travel. If they wanted to have rain they gathered a small number of stones, and clouds built up in the sky and rain would pour. If they desired to produce snow and severe cold they added stones, and it caused blizzards. It was reported that if they wished for rain or snow to be in another direction they used these stones. They pointed towards the district they wanted and then this region would be beaten by harsh weather. Indeed, this place would be struck by rain and cold. This is a complete and true story. Yet, this happened not due to their abilities or power but only because God wished it."

Abū al-ʿAbbās ʿĪsá b. Muḥammad b. ʿĪsá al-Marwazī continued his narrative: "Then I arrived at the city of Shāsh and met with a group of the local inhabitants. These people were familiar with stories about the Turks and have knowledge of their culture and manners. I asked them about this account and other stories and they told me that they had similar reports. The story by Bālqīq is the most accurate and precise, because he learned it from his ancestors and was well informed."

Abū al-ʿAbbās ʿĪsá b. Muḥammad b. ʿĪsá al-Marwazī added to this, saying: "In Shāsh I met with Ḥabīb b. ʿĪsá, he was an old clerk in the local administration and collected records about the history of Nūḥ b. Asad the Samanid (817–51) and his wars against the Turks.[67] He was well-versed in the affairs of this region. He handed me a copy of a letter that ʿAbd Allah Ibn Ṭāhir (d. 230/844) wrote to Nūḥ b. Asad. At the bottom of this dispatch was a fragment from a message that the caliph al-Maʾmūn had written to Ibn Ṭāhir. It contained orders to inquire about the Turks and explore the stories they narrate about the ability to cause rain. <642> Then Ḥabīb b. ʿĪsá said: 'Nūḥ b. Asad assembled the scholars of Shāsh and other well-informed Muslims of Turkic origin and asked them about these accounts. They agreed that these are true accounts but disagree about the reason for this power'."

Abū al-ʿAbbās ʿĪsá b. Muḥammad b. ʿĪsá al-Marwazī continued his narrative and added: "I heard Ismāʾīl b. Aḥmad the Samanid (892–907), who was the governor of Khorasan (Khurāsān), tell the following story: 'For several years I waged war against the Eurasian Turks. In one of these expeditions I rode at the head of an army of twenty thousand Muslims. About sixty thousand heavily armed Turks took to the battlefield against me. The fighting went on for several days. One day, while we were engaging the enemy, a group of Turkic slaves that served me, accompanied by some Turks from among those who were our collaborators, came to me and told me: "In the army of the infidel Turks that are facing us we have family members, brothers and other kin, and they have informed us that a sorcerer will come to their camp. They warned us and we are very worried by this news."

Ismāʾīl b. Aḥmad explained that the person that was named by his informants was a sort of Turkish diviner. The Turks believed that he had the power to produce clouds that would bring heavy rain, snow and winter weather; and that he was able to direct them against the enemy that he wished to destroy. Ismāʾīl's informants reported to him that the Turkic sorcerer decided that the Muslim camp would be demolished by heavy rain and terrible ice. Every creature stricken by this weather would die. Ismāʾīl drove his informants away, saying that although they had converted to Islam, in their hearts they were still infidels. He asked them: "Is a human being capable of making this story true?" They answered: "We have warned you and you know the odds. The event is set for tomorrow morning when the sun rises."

Ismāʾīl b. Aḥmad continued his story saying: "Our army camp was on the slopes of a mountain. Next morning as the sun was rising, a gigantic cloud developed on the top of this mountain. Continuously growing, it progressed towards our camp till it covered it completely. I became terrified by its darkness, by the scenes I saw and the noises I heard. It made horrifying sounds. Then I realized that it was a trial (*fitna*). I descended from my horse and prayed. The soldiers in the camp were like a rough sea, moving from side to side, up and down. They were sure that a terrible blow was going to hit them. I kneeled and prayed, calling on Allah to show his mercy. I covered my face with dust and begged God for help, praying: 'Save your believers O God. This trial (*miḥna*) makes your servants feeble. I know that you, Almighty, have the power, You alone control benefit and harm. O Lord, if this cloud pours rain on us it will cause heavy damage to the Muslims and strengthen the infidels. O Lord, remove this mischief from us. You only have the power and ability to do so, and only You are the Almighty and the Omnipotent.'"

Ismāʾīl b. Aḥmad continued his account: "Out of fear I prayed endlessly and did not stop from calling God. My face was covered with dust. I was certain that only God can provide us with good, only He can eliminate evil. While I was engaged in prayer, my servants and other soldiers rushed towards me with the tidings that the army was sound and secure. <643> They grabbed my arms and called on me to stand up and watch. I lifted my head and saw the cloud moving away from my army and advancing towards the camp of the infidel Turks. It started raining over them, pouring terrible cold on them. Like waves in storms they moved about. Their riding beasts broke away and their tents twisted. Those who were stricken by the weather became feeble or even died. My men said to me: 'Let us attack them'. But I said no. God is wiser and His punishment more bitter. Only a few of the infidels managed to escape. They left all their equipment in the camp and fled. Next morning we advanced towards the infidels' camp. We discovered there countless quantities of spoils. We carried the booty away and praised God, knowing that only due to Him were we able to win this decisive victory and to gain this immense booty. We thanked God for being safe and secure. Praise God the Lord of Creation."

A note on some Turkic cities and their marvels

Saʿīd b. Ḥasan of Samarqand narrated that several Turkic peoples are nomads. These groups erect their tents and take them down, and travel in search of grass for their livestock. In much similarity to the Bedouins within the Abode of Islam, they wander from one meadow to another to feed their animals. They are not subject to a king nor do they pay taxes to any authority. They attack each other and capture the women and children of other groups. It might happen that a clan would relinquish the territory where they were living and proceed to another territory carrying the captives with them. It might even occur that they would move to live next to a tribe whose women they had seized. They consider the children that these women bear their slaves. No one demands the release of these slaves. They have their own manners and customs (*sunna*) regarding them, which traditionally they agreed upon.

In the Turkic lands there are many cities. In some of these cities there are merchants and merchandises. In all of these cities there are bazaars. One of these cities is the city of the Toghuz-Oghuz. It is the biggest city in the Turkic lands. It is a well-protected town; no city has similar fortifications. It has a huge rampart built of rocks that is encircled by a moat filled with water. Its inhabitants are warlike and combative people. Swords are their main weapons.

ḤYUS is another city in the Turkic land. It lies not far from Shāsh. It too is a vast city. Its population does not adhere to any religion. They are the most evil creatures and no people equal their wickedness. Habitually they harass each other, the strong do not show mercy and would kill the feeblest. <644> Brothers do not trust each other, nor sons and fathers. They eat all sorts of animals. Prostitution is open. A person might enter the dwelling of another and sleep with his wife, while her husband watches them without becoming angered by this deed or voicing his disapproval of it. They are far from being brave, but are notably beautiful. Most of the men are effeminate and they drink blood. In the centre of the city is a lake rich with water. When one of them passes away, they throw his body into this lake.

DI is another city within the lands of the Turks. Its population does not fight; they surrender and pay taxes to those who seize their city. They have sexual intercourse with every creature that they meet be it a woman, a boy or an animal.

SWR is another city there [in Eurasia]. Its inhabitants fight against the people of Shāsh and Samarqand. They are very courageous and strong and capable of causing their opponents great harm. They are well trained in fighting. They are well practiced in the use of the lasso (*awhāq*) and use it in war and never miss the target. Among the male population of SWR there are many handsome men but most women are ugly. They disapprove of prostitution and kill those men and women who commit adultery. They have an alcoholic drink that they produce from various aromatics that they are familiar with. Drinking even a small quantity of it, of no more than one Baghdadian fluid ounce (*raṭl*),[68] causes heavy intoxication.

<645> JRISM is another city in the Turkic lands. Its inhabitants are busy mainly with attacking SWR. If they succeed in capturing a person from SWR they

will slaughter him, cook and eat him. They are savages; not even a single man among them is capable of understanding a simple word or possessing even limited knowledge. If a strong man from this city apprehends some one who is weaker, he might molest him. They too are very strong and courageous people.

AGHRS is another city in the land of the Turks. In their moderate temper and peaceful ways its inhabitants differ from all other Turkic groups. They eat most sorts of animals, but first slaughter them. They worship their own idols. Among this people are scholars that have profound knowledge of their tradition. Adultery is not seen. They refrain from performing disgraceful deeds. They have a large temple, which they claim came down from heaven in the shape that it is now. It is a high, wide and long structure. They explain that they pray to their particular idols in order to approach God, clarifying that "because the statues do not commit sins they would intervene on behalf of us." A king chosen from amongst them is their ruler.

KRSHIM (KMRSHIM) is another city in the Turkic lands. Its inhabitants eat every creature that moves on earth, without slaughtering or purifying it. They are endowed with valour and force. Men and women walk around naked and not ashamed, as if they were beasts. It happens that a man might meet a woman along the way and have sex with her while other people pass by and observe him. They carry out night attacks. In most of their fighting they use poisoned arrows. No government subdues them.

DKS is another city in the land of the Turks. They are powerful soldiers and able to fight long wars. Some of them are intensely caring for others. However, the rich are noxious to the poor. They have numerous sheep and horses and other cattle. Some amongst them are merchants that travel to the Abode of Islam. <646> They keep their promises even when they pledge for others who are not their neighbour, or even someone from another place. Nevertheless, they do not ban adultery. Their city is an affluent place. It is rich with water and there are plenty of gardens. They have various sorts of magnificent fruits that are not known in the Abode of Islam.

KYSAH is another town there [in Eurasia]. It is close to the land of the Khazars. The people of this town pressure the Khazars. They are the most evil and wicked creatures that God created. If a foreign person arrives in their town they will molest him. If they find a man with a lad, then this lad will become his servant forever. In this land there is a kind of animal that is the size of a dog. It likes to chase people and devours them. If it notices a person he has only slim chances of fleeing. This animal is very fast and would surpass a well-trained horse. The people of this land drink a very good and tasty white wine. It is delicious and has a very pleasant aroma. They eat dead bodies and consume blood, like lions. They have no mercy. They are malicious people with ugly faces and short bodies.

DANA[69] is another city in the Turkic lands. The men in this city are tall but the women are short. Their city lies between the <647> Khazars and Byzantium and the people of DANA fight both. They are strong enough to win battles against the first but are unable to beat the Byzantines. They eat all type of animals and lick their wounds [as dogs do]. In fighting they use swords but they cannot stand arrows. A single man among them might marry up to one hundred wives. It might happen that a husband would kill his wife and child in the presence of the victim's parents and siblings, and then they all would consume the woman's body. In the mountains of this region there is a sort of creature whose bites do not heal. Huge-tailed scorpions exist in their houses, they cause no harm and sometimes the dwellers eat them. In this place there are huge bats as big as doves and even larger.

SKUB is a city in the Turkic lands. Its peoples speak Syriac. They are a brave and courageous people. In skirmishes the wives participate along with their men and they fight vigorously. Prostitution constitutes a natural inclination of these women. A woman might look at a male and if she desires him then she will seize him. The man is unable to say a word. She will carry him to a mountain not far away from the city. In this mountain there are caves and grottos and she will place him in one of them. The poor man cannot escape. She will provide him with all the things that he needs. No one can gainsay her, neither her brother nor her child nor her husband – if she is married – can stop her. The man cannot gain his freedom nor be liberated from her grip, even if he is married and has children or a family. If he refuses to cooperate with her, she might kill him. This would be his fate if an obstacle prevents her from achieving her desire. On this occasion she would call other women to assist her. These women would help her to accomplish her objective because they have similar wishes and manners. They will fight together till success. When she gets bored of him or wearies of him or becomes fond of another man, she will release her captive and allow the man to return to his house. She will not allow other woman to meet in privacy with this man and would prevent others from taking hold of him, even when he is no longer with her.

<648> There is in the city of SKUB a marvellous hot spring. It bubbles from a cave in a high mountain and flows to ten houses in the city: three for women and seven for men. The houses are built from stone. In winter the water is very hot, but in summer the water's temperature decreases. There are in these mountains black, red and piebald foxes, however only a very small number of these foxes are hunted. This is so because these are adroit animals that hide between the trees and rarely descend to the plains. These foxes endure the bitter winter and cope with the snow in a way that no other animal does. Since snow falls in this mountain most parts of the year other animals that live there are also well equipped to survive the harsh weather.

There are additional cities that we will not mention and other locations that have not yet been visited by Muslims or other civilized people. We have no accurate information about these regions because they are far to the east and no person has reached them and no merchant travelled to these lands.

The titulary of the kings in the vast Inner Asian continent, in Khurāsān, the East, the lands of the Turks and neighbouring territories are:[70]

The king of Nīshāpūr	is Kiyān
The king of Marv	is Māhāyah
The king of Sarakhs	is Rādhūwiya [Zadhuwiya]
The king of Abīward[71]	is Bahamna
The king of Nasá[72]	is Irān [Ibraz]
The king of Ghuristān	is Barāz Bandeh
The king of Marv al-Rūdh	is Kīlān
The king of Zabūlistān	is Fīrūz [Fayrūz]
The king of Kābul	is Kābul Shāh
The king of Tirmidh[73]	is Tirmidh Shāh
<649> The king of Bāmiyān	is Shīr Bāmiyān
The king of Sughd	is Ikhshīd[74]
The king of Farghāna	is also Ikhshīd
The king of Rūyshān [Rubshan]	is Ruyshān [Rubshān] Shāh
The king of Jūzjān	is Guzgan Khodāh
The king of Khwārizm	is Khwārizm Shāh
The king of Ḥabsh [Khans]	is Ḥabsh Gīlān
The king of Bukhāra	is Bukhārá Khodāh
The king of Ushrūsana	is Afshīn
The king of Samarqand	is Tarkhūn
The king of Sijistān and al-Dāwar	is Zunbīl [Rutbīl]
The king of Herat, Būshanj and Būdhghīsh	is Barāzān
The king of Kush and Rakhj	is Nīrūn
The king of Transoxania	is Shār Shāh

The kings of the Turks are Haylūp Khāqān (Xaqan), Jabghūn, Shābih Khāqān, Sanjabr [Sinjibu] Khāqān, Māyūs [Manush] Khāqān and Fayrūz Khāqān. The minor Turkic titles are: Tarakhān, Nīzak, Jūr-Tekīn, Yamrūn, Sahrān and Ghūrik.

[Plato (*Aflaṭun*) said: Among the Turks sincerity is unfamiliar, among the Byzantines generosity is unknown, amongst the Khazars shame is unrecognized, sadness is alien to the East Africans, amongst the Slavs bravery is unknown and amongst the people of the Indus Valley continence is strange.]

This brings our book to its end. Thanks be to God that led us up to here. Aḥmad b. Muḥammad b. Isḥāq al-Hamadhānī, who is known as Ibn al-Faqīh, [wrote] this *Akhbār al-Buldān* (The Book of the Lands). Praise God, the Lord of the Universe. May God bless our prophet Muḥammad, the Seal of the Prophets and the Messenger of God.

c) Abū Dulaf: Pseudo-travel

The ʿAbbāsid bibliographer Ibn al-Nadīm (d. 995) says, in a chapter describing the sanctuaries of India, "the globe-trotter (*jawwāla*) Abū Dulaf al-Yanbūʿī told me".[75] The famous *littérateur* Abū Manṣūr al-Thaʿālibī al-Naysabūrī (961–1038) supplies a biography of Abū Dulaf. The meagre information enables us to reconstruct the period, social circles and works of this tenth-century man of the pen. The manuscript of Abū Dulaf was discovered in Mashhad (Iran) in the 1920s.[76] It is a pseudo-travelogue of a journey that never occurred, which was composed as an epistle (*risāla*); that is, a written informative message. Modern scholarship has played down the value of Abū Dulaf's account, describing it as a totally bogus account replete with unlikely travel details.[77] Nevertheless, it was well received by some Muslim authors who wrote about Central Asia (Turkistan). Moreover, its popularity provides the account with additional value, since it serves as a solid tool for depicting the imaginary *mappa mundi* of the medieval Arab reading audiences. These combined features make an English translation worthwhile.[78]

The first epistle sent to us by Abū Dulaf Misʿar b. al-Muhalhil al-Yanbūʿī, mentioning in it his observations and adventures in the lands of the Turks, China and India. Sir, I your humble servant have witnessed, O my lord, your enthusiastic interest in writings and keen attention in composition. Hence I decided, may the Lord prolong your life, not to let your expectations down and not to fail in addressing your wisdom. I decided to share with you some of the experiences and enterprises I have encountered in the past, aiming to clarify rumours that were heard and to give fully what the heart needs.

Thanks to God and blessings upon His prophets. I will open with an account of the roads in the east and the variety of governments in those countries, explaining the rulers, the changing conditions, the different religious rituals, the haughtiness of their kings, the urbanity of their political principles, the structure of their governmental systems. The knowledge of these adds to insight and is a must in personal tutoring. Allah spurs wise and prudent persons on to enquiring into it and encourages sensible and practical minds to learn it.

He said in the Holy Qurʾān: "Have they not journeyed in the land and observed how the final punishment of those who preceded them was? They were stronger in might and superior than themselves, they tilled up the earth and cultivated it more than they themselves have populated it; Messengers came to them with the clear warning signs; and God would never wrong them, but they wronged themselves."[79] He made learning from experiences a duty and deduction from the remote past an obligation. I do not suspect that you will go astray along the road that leads to rewards and I am sure that you hope for sound advice. Due to the strong bounds of affection that connect us and as a hearty indication of my submission, I found it necessary give this information to you.

I pray for God's help and hope to refrain from fault. He is the Patron who helps. He is the All-Mighty, the All-Wise.[80]

Crossing considerable space and distance from my country, I continued eastward till I reached to the land of Khorasan (Khurāsān). I observed the governor Naṣr b. Aḥmad b. Ismāʿīl b. Aḥmad the Samanid (914–943), who was nominated to administer the province.[81] He is an efficient ruler, a most able administrator, who dwarfs [other] talented administrators and for whom even efficient and brilliant officers are not a match. I met in his court with the emissaries of Falin (or Qalin)[82] b. al-Shakhir, the king of China.[83] They expressed their king's wish to marry Naṣr's daughter and willingness to establish marital connections with him. Naṣr refused, arguing that the Islamic holy law forbids it. In response they suggested that his son marry the daughter of the king of China. To this suggestion Naṣr agreed. I seized the opportunity and travelled with them to China. On our way we passed across the land of the Turks.

The first tribe we encountered after traversing the Islamic territories in Central Asia was the Kharka [*kharga*= yurt]. It took us a month to cross their region. During this period we consumed wheat and barley.

Then we entered the land of the Takhtakh tribe. We consumed barley and millet (*dukhn*),[84] various kinds of meat and prairie vegetables. For twenty days we travelled in security and calm. They observe the king of China's orders and obey him. Because of their closeness to Islam, and even conversion, they pay tributes to the Kharka and support them in most cases of raids on neighbouring polytheists.

We arrived at a tribe known as al-Baja. In their territory we consumed millet, lentils and chickpeas. For several months we travelled in their territory in security, although they are polytheists who pay taxes to the Takhtakh and adore their kings. They pray to the cow. Because they worship cows they do not possess them or own them. The land is rich with many grapes, figs, and black azarole. There also is a sort of tree that does not catch fire. From this wood they produce idols.

Next we crossed to the territory of the Bajānāk (Pečenegs).[85] Their faces are covered with long beards. Their customs are barbaric; they frequently engage in violence. A man might sleep with a woman on a public road. They eat only millet. For twelve days we travelled in their territory. They told us that their territories stretched far to the north and border with the Slavs. They do not pay taxes to any power.

We continued and travelled to a tribe named Jikil (Čikil).[86] They eat barley, grass peas and mutton only. They do not slaughter camels and do not acquire beeves. They wear only wool and furs. A small number among them are Christians. They have handsome smiling faces. A man among them would marry his daughter, sister and other related women that Islamic law forbids. They are not Zoroastrians, although their marriage practices are akin.

They worship stars and pray to Canopus, Saturn, Orion, the Bear (Ursa Minor) and Capricorn. [The star] Sirius is named by them the Supreme Lord.[87] Their temperament is mild and they do not demonstrate wickedness. All their neighboring Turkish tribes are eager to grab hold of them and kidnap from among them. In their territory a plant known as leek (*kilikan*) grows, it is very tasty and often is cooked with meat. In their lands are pits of bezoars and herds of Chinese cattle, which is a sort of animal that lives in the steppes [and its horns are used]. From blood and tar [produced from wild ingredients] they produce a heavy alcoholic drink that easily intoxicates. They construct their dwellings from wood and bones. They have no king. We travelled for forty days in their lands in security but they were days of scarcity.

Next we entered the territory of the Baghraj tribe. They grow moustaches but not beards. Their riders and infantry are well trained in the use of spears and maneuver with them magnificently. They are ruled by a powerful king who is said to be a descendant of Yaḥyá b. Zayd, the grandson of ʿAlī b. Abī Ṭālib (d. 743). He possesses a gilded codex of the Holy Qurʾān. Eulogizing verses are written on its back, lamenting the late Zayd b. ʿAlī b. Ḥusayn, the grandson of the Prophet Muḥammad. They adore this codex and worship it, considering Zayd to be the king of the Arabs and ʿAlī b. Ṭālib to be the God of the Arabs. They do not accept the kingship of a person who is not of the progeny of Zayd b. ʿAlī. While they turn in prayer to the sky they open their mouths and raise their eyes saying that the God of the Arabs will descend from the sky and ascend. Zayd b. ʿAlī's progeny who lead them have long beards, protuberant noses and wide eyes.

The Baghraj tribe consumes millet and mutton – only of rams. There are no cattle or goats. They wear only felt and do not use any other kind of cloth. They believe that serpents should be burned. For a month we travelled in their land in great fear and confusion. We paid them a tax of ten percent on the goods we carried.

Next we entered the territory of the Tibet tribe. We travelled for forty days in their country, in ease and security. Their nourishment is based on wheat, barley, beans and meat and fish, vegetables, grapes and fruits. They dress in a variety of clothes. In their country is a large city built from reeds. In this city is a temple made from lubricated skins. In it are idols made from cattle horns and the horns of musk deer. In this country live communities of Jews, Muslims, Christians, Zoroastrians and Buddhists. They pay taxes to the king of the Baghraj from the house of Zayd b. ʿAlī. They elect their king in a lottery. They use jails to detain criminals and wrongdoers. They pray towards the direction we pray, i.e. Mecca.

We continued to the land of the Kīmāk.[88] They make their home from animals' hides. They eat chickpeas, Egyptian beans and flesh. They consume the meat of male sheep and goats but forbid the consumption of ewes or she-goats. In their country grows a kind of grapes that carries unique fruits, half of the berry is white and the other half is black. In their land is a particular stone known as the magnetic stone. Whenever they desire they use this lodestone to cause rain. In a plain in their country are goldmines from which they carry chunks of the metal. There are also diamonds that water torrents expose, and delicious plants, some that narcotise and some that keep vigilant. They do not have a king nor a shrine. They worship the aged that are more than eighty years old, unless there is a visible defect in their body or a flaw. We travelled for thirty-five days in their country.

We arrived at the land of the Ghuzz tribe. They have a city built from stones, wood and reeds. There is a temple but no idols are in it. They are ruled by a supreme king who collects taxes. They trade with India and China. They only consume wheat, because there are no vegetables in their lands. They eat the flesh of sheep and goats with no gender taboo. Their dress is made from cotton and fur robes. They do not wear wool. In their land are found white stones that serve as a remedy for colic and a red stone that blunts swords if they are smeared with it. We travelled a month in their country, in peace and security.

Then we reached the land of the Toghuz-Ghuz tribe. This people eat slaughtered and non-slaughtered meat. They wear cotton and felt clothes. There is no temple in their land. They worship horses and treat them with great care. A stone in their land stops haemorrhages and bleeding. They pray to the west and their banners are black. For twenty days we travelled in their land in enormous fear and anxiety.

Next we arrived at the land of the tribe named Khirkhīz (Kirgiz).[89] They eat rice and millet, and the flesh of cattle, sheep and goats, and other kinds of meat except for camels. They have a temple. In it is a pen with which they write. They are wise and mindful people. They do not extinguish the light of burning lamps and leave them blazing till the flammable substance dies out. During prayer they speak in rhymes. The musk in their land is very limited. During the year they feast on three holydays. They carry green flags. Praying towards the south, they adore Saturn and Venus. Mars troubles them. They worship newborn infants and serve the pregnant mother till she gives birth. They wear a variety of clothes with no restrictions. Busy in trade, they use clocks and calendars. A considerable number of lions exist in their country. They employ a shining stone and favor it over lanterns. This stone can be found only in their country. They are governed by a learned king who reads what they write to him. They obey him. No one younger than forty years of age is permitted to sit in front of the king. We travelled in their territory for a month in security and comfort.

We continued to the country of the Kharlukh (Qarluq) tribes. Their nourishment consists of chickpeas and lentils. From millet they produce an alcoholic drink. They eat salted meat and no other style. Dressed in woolen garments, they worship in a temple decorated with pictures of their past kings. It is built from incombustible wood. These trees grow widely in their land. Oppression and infringement are apparently observable. They often assault and fight amongst themselves. Fornication is very widespread and not considered to be forbidden. Gambling is their manner of deciding. They bet on their mothers, daughters, sons and wives. Their gamble continues till the loser gives up and buys off the winner, who takes the prizes and sells them to a merchant, if he wishes so.

Their women are outstandingly beautiful. Corruption is very common. Jealousy is very weak amongst them and they hardly know any sexual taboos. A wife of a leader or other man, his daughter or sister, would approach an arriving caravan and scrutinise the faces. If she likes one of the travellers she would invite him to her house and accommodate him. She would employ her husband, brother or son to care for the guest's needs. As long as she gives the visitor lodging, her husband will refrain from relations with her. He will contact her only to accomplish her wishes and then he will withdraw. While the woman and the man she has chosen enjoy themselves, eat and drink even as the husband is watching, he does not demonstrate any signs of resentment or express any objection.

They have a festival during which they have to be dressed in brocade (*dibāj*).[90] Those who do not obtain this expensive textile settle for embroidering a piece of *dibāj* on their dress. They obtain silver ore and from it they extract quicksilver (mercury). In their country grows a tree resembling myrobalan (*ihlilaj*). If a hot swelling is coated with its sap it heals immediately. They worship a huge rock and gather near it to take decisions. The stone lies on its back, its colour is green and they sacrifice offerings to it. We travelled in security in their country for twenty-five days.

Continuing our road we entered to the territory of the Khutlukh.[91] We traveled amongst them for ten days. They only consume wheat and uncooked meat. I did not experience among the Turks a tribe more violent then the Khutlukh. They frequently kidnap their neighbours, yet those who mingle with them will enjoy safety. The marriage of sisters is a common practice among them. A woman marries only once in her life. Widowed, she will not remarry. They are wise and capable. Fornicators are punished harshly; the man and the woman are burned to death. Divorce is not practiced among them. The bride price is the whole property of the husband, and a year serving her guardian. They permit revenge for bloodshed. A man who harms another person will compensate the wounded. However, if the injured man were to perish after the collection of the compensation payment, his blood will not be avenged. Their king abhors evil. Their king may not marry and if he marries he will face execution.

We carried on our journey and arrived at the country of the Khitan tribe.[92] They consume barley and vetch (*Lathyrus spp.*) and roasted flesh only. Their marriage practices are correct, their jurisdiction sensible and rational. They are not ruled by a king. Each group of ten is guided by a prudent and shrewd elder. In cases of jurisdiction they ask his opinion and agree to his verdict. They do not punish abuse harshly nor murder. They have a temple and confine themselves to this edifice for the period of a month, or for shorter or longer periods. They do not wear pigmented textiles. In their country are excellent musk deer, yet this quality of the musk is preserved [only] as long as it is kept there, if the musk is carried to another land its substance changes and alters. They grow a considerable number of diverse vegetables, which generally are very valuable. A sort of viper in their mountains kills whoever looks on him. These vipers never depart from this mountain. They have a stone that helps in cooling down fibre; it affects only their land. There is also high-quality bezoar resembling wax with green veins. We travelled for twenty days in their land.

From there we continued to a city named Biha(?). In this place grow many palm trees, plenty of vegetables and grapes. There is an urban centre surrounded by villages. It is an administrative centre controlling nomadic and sedentary populations. Their king bears the title Biha. In the urban centre dwell various communities: Muslims, Jews, Christians, Zoroastrians, Buddhists and polytheists. They celebrate several festivals. They have a green stone that serves as a remedy against ophthalmia, and a red stone that is useful in treating inflammation of the spleen. They grow three kinds of high-quality indigo plants: *qani* (crimson), *tafi* (floating) and *murtfa*ʿ (rising). Textiles dyed with it do not sink in water. We travelled across these lands for forty days, ranging between tranquility and fear.

Continuing we reached Qulayb (or Qalīb?).[93] In this land live Arab nomads (Bedouins) who were confederated with one of the kings of southern Yemen

(*Tubba'*) who raided China and arrived in this land with his army.[94] They have summer and winter camps and palm plantations in various regions of sand and water. They speak ancient Arabic, and do not use another tongue. Their script is the ancient royal script of southern Yemen (Himyarite)[95] and no other writing. They worship idols. Their king is a scion of one of the aristocratic families. The nominated king has to be from this royal family and cannot be the offspring of another house. They have regulations and restrictions regarding fornication and adultery. From dates they distil an excellent alcoholic drink. Their king regularly sends gifts to the king of China. For a month we travelled in their land in fear and constant danger.

Then we arrived at the Station of the Gate (*maqām al-bāb*). This is a city built from sand that houses a guard of the king of China. Those Turkic tribes that wish to enter his territory send their request from there, asking the king's permission to enter China. We spent three days there as guests of the king. At every parasang we were provided with a fresh riding animal. We stopped at the wadi of the Station and asked permission to enter his kingdom. The messengers advanced ahead us. After a period of waiting in this wadi, the authorization reached us. This is a lovely wadi, one of the most pleasant and beautiful places on earth. There we spent three days as the guests of the king. Receiving the authorization, we travelled across the wadi for a long time before seeing the city of Sandabil, the capital city of China. In that city is the royal palace. We encamped in a distance of one stage (*marḥala*)[96] from the capital.

Next morning we continued our journey. We travelled all day long and reached the city at sunset. It is an immense city. To cross it takes the whole day. There are sixty main avenues. Each of them leads to a gate of the royal palace. The walls rise to the height of ninety cubits and this is also their width. On the walls flows a tremendous river, which splits into sixty conduits. Each canal leads to a gate of the royal palace, where a water wheel distributes the water to smaller channels that flow into yet smaller ones till the water is absorbed into the ground. Half of the water gushes under the wall and irrigates the gardens. The other half returns to the city and is utilized by the inhabitants of the streets outside the royal palace. In each street are two canals. Each toilet room has two pipes that are arranged so that they are opposite each other. The incoming stream waters the population while the outgoing stream carries away their sewage.

They have a gigantic temple, which is said to be bigger than the al-Aqṣá mosque in Jerusalem. It contains sculptures, pictures, idols and a large statue of BDH [Buddha?]. They conduct skilful politics and sophisticated jurisprudence. They do not slaughter animals nor consume flesh. If a person kills an animal he will face execution. The territory of China is [larger] than the combined size of the kingdom of India and the land of the Turks.

I was received by the king of China, noticing his high competence, sagacity and intelligence. The emissaries presented him with the scheme that his daughter would marry the son of Nūḥ b. Naṣr, and the king agreed to the proposal. He showed great respect towards me and the rest of the delegation. We were his guests till his daughter's preparation was completed and she was ready to depart.

He sent two hundred eunuchs, and three hundred top servant girls that were selected from his private entourage. We bore the bride and the retainers and brought them to Khorasan, to Nūḥ b. Naṣr. The Chinese princess gave birth to ʿAbd Allāh b. Nūḥ. Naṣr b. Aḥmad died prior to her arrival in Bukhara. His realm passed into the hand of his son Nūḥ b. Naṣr who benefited from her.

2 EURASIAN TURKS AND THE LEGITIMATION OF THE SALJŪQS

a) Ibn al-Dawādārī: A Turkic creation myth

Abū Bakr b. ʿAbd Allāh was the grandson of a Turk emir in the service of the Ayyūbids and the son of a high-ranking Mamlūk administrator, who was raised in Cairo and Damascus. He held some official post during the reign of al-Malik al-Nāṣir Muḥammad (first third of the fourteenth century).[97] Two of his works have survived: a chronicle from the *Creation to the Mamlūk Sultanate* in nine volumes and an epitome.[98] The ancient Turkic creation mythology,[99] which is translated here, is presented as an ego-document in both accounts.

I [Ibn al-Dawādārī] tell an amazing story and an incredible account regarding this people (i.e. the Tatars or Mongols), their emergence and their early history. Presumably no other historian has mentioned the story, because no one has ever heard it. It happened merely accidently that I stumbled upon a marvelous book that the Turks highly adore. In Turkish the book is named *ulū ay atām bitkī*, which is translated [into Arabic] "the book of the superb father" (literally: our supreme moon father has given).[100] By chance I came across it in 710/1310–1311 when a man named Amīn al-Dīn al-Ḥamawī handed me the book for consultation. This man served as a secretary of the late emir Badr al-Dīn Baysarī.[101] My late father served during that time as the director of the Eastern Provinces (i.e., east of the Nile) and the adjacent districts. This Amīn al-Dīn had a secure paid position in these provinces, because of this he often called at the town of Bilbays (Bilbeis). Since at this time I was young, I was eager to learn about various dimensions of praise literature, *belles-lettres* and other genres of writing. The abovementioned Amīn al-Dawla was an excellent companion and narrator, the soirées in his company were the most pleasant, the most abundant in refined manners and culturally the most illuminative. He frequently visited leading personalities and nobilities of this age. Among those he met with were the sheikh Jamāl al-Dīn al-Samlūṭī, the physician Shams al-Dīn [Muḥammad] ibn Dāniyāl the oculist and writer of shadow plays (*ṭayf al-khayāl*),[102] Jamāl al-Dīn al-Balālīqī (a satirist or popular poet) known as Ibn Zaytūn who composed nice popular poetry (*balāliq*),[103] a man named Manṣūr al-Adīb al-ʿAbbāsī and other dignitaries and literati. With refinement and etiquette we forgot about daily concerns, extracting top qualities from perfect men in every topic and subject.

On that particular day we discussed and agreed upon the art of history and talked about the coming of the Tatars (Mongols). The abovementioned Amīn al-Dīn noted that he owns a book that no one else posseses, there is no match to it in the Abode of Islam. This book had been the property of Badr al-Dīn Baysarī, who considered it to be his most exceptional and priceless possession. Whenever he brought the book out he demonstrated his deep affection for it, carrying it on his head and displaying his admiration as if it were the Holy Qur'ān, the Book of Allāh. So we asked Amīn al-Dīn to produce the book so we could examine it. He departed and returned with the book. We inspected it and noticed that it was a valuable book of the utmost price. The script was fine, resembling the handwriting of [ʿAlī b. Hilāl] Ibn al-Bawwāb's students.[104] Its paper was manufactured in Baghdad (*waraq baghdādī*).[105] The outside of the book's cover was red Atlas satin,[106] on its inside was yellow [golden silk]. The padlock on it was made from gold. [I never saw anything like it.] These demonstrated that great care was taken to guard the book.

We assembled around Amīn al-Dīn and he read to us from the book. [Only Ibn Dāniyāl was missing.] We figured out that it is a history of the ancient Turks and their origins, according to their creed that contradicts the unstained holy Islamic beliefs. Since I discovered in it wonderful and incredible *mirabilia*, I found it suitable to transcribe from it the section dealing with the myth of origin of the Tatars (Mongols) and Turks and their antecedents. Because the amazing stories are not in line with the tenets of Islam, the legends – which are not true and reliable tales – aim to amuse. These speculations were presented by philosophising sages who intended to capture the minds of kings in whose service they worked. These kings' ambitions were the prolongation of their worldly lives. These philosophers' essential belief was that the natural span of life is 120 years. Any person that passed away at a younger age had certainly neglected his soul and failed to comply with stipulations that they had laid down. Thank God, we refrain from their erroneous suppositions. He shelters us from their rhetorical flourishes and empty words. We know and adhere to the belief that the span of life is predetermined. When death's date comes God delays it not. This is the sound faith and the excellent wit.

In order to confront the hypotheses that these Tatar/Mongol people present, I will put them forward in the following chapter and note down passages copied from the abovementioned book. This will make clear their erroneous ideas, which they incorporate in fairy tales.

Jibrīl (Gabriel) b. Bakhtishūʿ (Bokhtishoʿ), the slave of God who relies on his much-forgiving and pardoning Lord, in the year 211/826–7 rendered this book from Persian to Arabic, which earlier had been translated to Persian from Turkish. Following his working method of leaving Turkish expressions intact in the Persian translation I also left several original expressions in Turkish and did not translate them. This was also the working method of Abū Muslim, the mastermind of the ʿAbbāsid cause, who had transmitted the original Turkish myth to Persian. Jibrīl b. Bakhtishūʿ said that this book was among the valuable belongings that had reached him from [the hoards of] Abū Muslim al-Khurasānī (d. 755), the mastermind of the ʿAbbāsid propaganda machinery (*daʿwa*).[107] He also said that the book was in the library of Buzurkmihr [Brzjmahr in the printed text; read Bozorgmihr or Bozorgmehr] b. Bukhtakan the Persian, the alleged ancestor of Abū Muslim,[108] who claimed to be a descendant of this famous counselor.[109] Because he was blessed with a sound mind and demonstrated high managerial and political qualities, as well as wide-reaching knowledge in a variety of topics, this claim of his might be not false. Al-Manṣūr, who was troubled by Abū Muslim's wisdom and professionalism, executed him.[110] I (Jibrīl b. Bakhtishūʿ) heard the caliph al-Maʾmūn saying: "Three rulers governed the land well, handled state affairs skilfully and tutored dynasties: Alexander, Azdashīr [read Ardashīr] and Abū Muslim." Jibrīl b. Bakhtishūʿ augmented these lines with effusive praises of Abū Muslim, yet since it is beyond our needs I skipped them and did not copy these passages.

Jibrīl continued his narration [and cited several chapters from that book]: In the remote reaches of upper China, in the direction of the rising sun, Allah the Almighty, may His strength grow and His wisdom increase, created a towering mountain which soars to the height of four parasangs. This mountain rises straight up to heaven. It is made from massive black stone and no plants grow on it. Allah, whose unrestricted knowledge man cannot fully comprehend, made it as a barrier to protect the earth from the black ocean and as a fence to protect from the rising sun. Allah's concealed wisdom and His incomparable acumen made the sun rise from the depth of this black ocean. The mountain blocks the sun's rays from striking the earth in the first moment of dawn so that the morning heat will not burn the earth. This is Allah's grace and the conduct of God's wisdom, because if the sun as it rises early in the morning was to overheat the earth, not a single creature would survive on its face, no plant would grow. Hence the sun's rays reach the ground only twelve minutes after daybreak in the east, when it is high in the sky and the heat is less. This benefits all. In Turkish the name of this mountain is Qarāṭāghjī (*ulu kara dağ*) which means the [Great][111] Black Mountain.

From the western slopes of this mountain several springs of sweet water gush out. The water is sweeter than honey, cooler than snow and has a good odour stronger than perfume. These springs flow to a lake whose full diameter is seventy parasangs. From this lake a river emerges. The river runs parallel to the mountain's slopes, that is to say a distance of 170 parasangs. In the middle of this distance lie two great cities.[112] Walls built from black stones encircle these two cities. The stones are carved so that each stone is joined to its neighbour, tongue and groove (literally: male and female).[113] It is so smooth that the interlocking is almost impossible to see. One has to inspect carefully to detect a cut. The length of each wall is seventeen parasangs. Each city has forty gates made from Chinese iron, which resembles ornate silver. Each gate is protected by a towering bastion made from these materials. The river cuts across these two cities, which are separated by a distance of seventeen parasangs. The river's water is allocated in a most sophisticated and complicated system so that each of the houses in these twin cities is supplied with water from the river. The river flows out from the cities, splits into several branches and waters valleys in that region, providing rich farms, fields and orchards that yield magnificent crops and delicious fruits. Only God knows the marvel of it. The river system encompasses districts and populated regions that take a year to cross, either from south to north or from east to west. They are inhabited by diverse peoples and nations with different characters [and religions]. Yet all are the subjects of the ruler of this pair of cities.

The paired cities are named in Turkish Aydarmāq and Asharmāq. All their rulers are the offspring of a single common father. Every king who rules the cities is named in Turkish Alṭun (Altın), which means the gold king. The population is neither threatened by enemies who frighten them nor by adversaries who terrify them. They live affluent and enjoyable lives, agreeable and pleasant, eating, drinking and procreating. They have the most beautiful faces and youthful bodies and live pleasurably. They consume high-quality meat and sweet fruits that the rich soil nurtures and the outstanding meadows produce. Their fields are covered with a variety of grasses that grow in these fertile places where the air is fresh and the water clean. For these reason they are unfamiliar both with sickness and with deformations. Their hearts are governed by joy. Even if death occurs and a person passes away they do not mourn him nor lament. They enjoy long lives. No one dies before reaching the age of a hundred or even older.

[Regarding this they hold to detailed and exhaustive beliefs, which resemble the essence of the opinions of the philosophers and the arguments of the thinkers. We refrained from providing a detailed account of these principles because they are not in line with the sacred Islamic religion. They actually resemble the opinions of those who claim that the natural span of life is 120 years. They assume that it is possible to foresee from the structure of the body the prospects of life expectancy and the date of man's death. They claim that the originator of this method was the ancestor of this Turkish people whom they call *ulū ay aṭājī*, which means in Turkish (*ulu ay atacı*) the Great Sovereign Moon Father.[114] They have amazing legends about his creation that we refrain from narrating because there are not in line with the Holy Islamic Law.]

Jibrīl b. Bakhtishū' said: Regarding this (patriarch) they have detailed and telling legends, which are at the heart of our account of him. Bozorgmehr b. Bukhtakan the Persian, the author of the book we depend upon, says: Although his words are not in line with the true religion of Islam, yet we quote him not to validate his arguments but only for the sake of amusement, and the transmitter of a historical narrative should not be blamed. He says that the first creatures (to inhabit Earth) were on this Qarāṭāgh (*Kara dağ*) Mountain. High on this (black) mountain there is a lofty cave. It will take three long days for a shrewd and diligent man to climb along a difficult, winding and rocky path until he reaches it, exhausted after a wearying experience from bottom to top. At the mouth of this cave is a huge gate of red gold, decorated with various large precious stones. Guards and servants administer the place. The cave is a shrine worshiped by these regions' population. If one among them wishes to take upon himself an obligation he points towards the cave and swears: "*Ayyama Qarāṭāgh.*" The meaning of these words is: "In the Name of the Black Mountain."

Jibrīl b. Bakhtishūʿ continued quoting Bozorgmehr b. Bukhtakan and narrated the principles of the Turks' belief system. We are transmitting his words not because the validity of this wrong faith is accepted by us, but rather to inform. In due course we shall challenge their superstitions, using both sound reasoning and the true tenets of Islam.

Jibrīl said: During the early stage of history heavy rains caused torrents that carried mud to that cave, which filled up a furrow moulded in a shape resembling the human figure. At that time the sun was in the zodiac sign (*burj*) of *ufnūn*,[115] which in our terminology is the Crab (*saraṭān*: Cancer; June 21–July 22).[116] Jibrīl means that the sun was at its zenith, radiating directly and causing the highest temperatures that this geographical latitude gets. Each geographical region has its climate according to its latitude and the azimuth of the sun. The rest of the liquid that flowed down and burst through the cave came together in this furrow. The storm calmed and the sun baked the cave, which resembled a woman's womb (*jawf*). The substance that filled up this furrow was made from two elements: water and earth. The sun at zenith was as hot as fire and baked these two elements. Then during a period of nine months mild winds cooled it. The four elements fused and became fully homogenised. Nine months passed and the sun was in the zodiac sign of *ḥītān*,[117] which in our terminology is the Ram (*al-ḥamal*: Aries; March 21–April 20). Three months constitute a season. As it happened the sun was at zenith, which is the beginning of summer. Since nine months had passed, that is to say the summer, autumn and winter seasons, the sun reached the temperate season. The sun was in the focal point and the temperatures were mild. The substance reached its ripening, perfection, and full size while the sun was not too hot or too cold. The moment of emerging arrived and creatures in the form of speaking serpents came out from the furrow.

Jibrīl said that this account is accepted as true by those who consider decomposition as a truth (*ahl al-taʿāfīn*, i.e. emanationism).[118] They prove it by claiming that every organ produces a living spirit during its decomposition, and this is by necessity, as is shown by in worms in vinegar and cheese or insects in earth. Often it happens that various creatures come out from rotten organic materials.

He [Jibrīl] adds that the person that emerged from this cave is named in Turkish *Ay Aṭām* (*Ay Ata*), which should be translated into Arabic *al-ab al-qamar* (Father Moon).[119] The Turks claim that they are more entitled to see him as their ancestor, while the Persians maintain that they have the priority. The Persians call him K.H.U.M.R.T. (Gayomart or Kayumart; Eternal Life), which in Persian means the father of the fathers.

I say: We Muslims pray to find shelter with God and to abstain from agreeing with the Turks' cosmogony or the Persians' faith. We proclaim that Allah created Adam from dust; He created man by His exclusive and incomparable power. The Prophet Muhammad transmitted this narrative. He is the Truth and all his sayings are true. Allah created everything *ex nihilo*. The false claims of the Turks and Persians that are narrated by the composer of the mentioned Turkish book are indeed extremely ridiculous. Neither a sound mind nor the Islamic faith will accept them as valid tales.

Jibrīl b. Bakhtishūʿ continued quoting Bozorgmehr b. Bukhtakan: That man, who is named [in Turkish] *Ay Aṭām* (*Ay Ata*: Father Moon), fell to earth, to a place with fresh air and clean water and lived there for forty years. He accumulated strength and developed, his organs grew and his intelligence increased. Then the water amassed once again and torrents washed the earth. The results were similar to the early flood. Yet at that time the sun was at the *k.y.nān* sign in the zodiac, which we should assume equals the *sunbula* (Virgo, or *ʿadhrāʾ* in Arabic; 23 August–22 September) in our zodiac, and the burning sun, which was descending at the western horizon, baked the earth. This was fully ripened when the sun was high in the eastern sky. Then a female figure emerged.

This is a shortened version of the book's long tale. Were it possible I would prefer to cut this saga short, yet the obligation to present an eloquent account of the emergence of the Tatars (Mongols) compels me to prolong. I came not across a single book that speaks about the early history of the Mongols, all other historians start with the history of Genghis Khan Temujin [*jik.z* (*sic*); read *Jinkīz khān tamirjī*] and do not overstep the bounds of this chapter. Yet I found in this book an account of the Mongols' early history and I believe it is in place to transmit it.

Jibrīl b. Bakhtishūʿ continued and narrated: This figure emerged as a feminine figure because she was not created during the period when the sun reaches the highest point in the sky. Falling short of the hot temperature, she lacked male organs. She was called in Turkish *Ay Wā,* which means *wajh al-qamar* in Arabic (Moon Face). Afterwards the man and the woman married and procreated. Unchangeable Ay Aṭām (Ay Ata) survived forty years after the descending of Ay Wā. Then for forty years he declined and became smaller. When he reached the age of 120 years he passed away. His eldest son carried the body back to the cave and entombed Ay Aṭām (Father Moon) in the furrow, hoping that his father would revive. Ay Wā died forty years after the death of Ay Aṭām. Her son carried his mother's body to the tomb of his father. He discovered that Ay Aṭām's corpse had disintegrated and placed his mother's body upon him. He buried them and left the place. On the cave's mouth he built a golden door and nominated guards to protect the cave.

Jibrīl b. Bakhtishuʿ said:[120] "The progeny of this man increased and augmented. Some did wrong to others.[121] They agreed to pick one man who would serve as their leader. To him they would address all questions and he would judge cases of violence. They concurred that the eldest son of Ay Aṭām would be their king. The first among the Turks to sit on the throne was the eldest son of Ay Aṭām. In Turkish he was called Ay Aṭām K.SH.K.R.I [read *kujukrī*, i.e. *küçük* in modern Turkish],[122] which in Arabic is *al-ab al-qamar al-ṣaghīr*, that is to say Moon Father Junior (or the small Moon Father). For eighty years he was the king. He is the person who built these two cities, Aydrmāq and Ashrmāq. It took him forty years to construct the paired cities. After Ay Aṭām Kujukrī's death he was succeeded by his son Kujukrī B.L.J.K.I (i.e. *piliç* in modern Turkish)[123] that should be interpreted in Arabic as Kujukrī Farkh (young or chick) or Little Young Bird. Before his death Ay Aṭām Kujukrī had ordered his son Kujukrī Bilijkī to produce an effigy of his corpse from solid gold and to sit this mummy on a throne in a temple that he would construct. Kujukrī Bilijkī accomplished his father's will. He built a highly structured temple for his father. The flames of golden oil lamps, supplied with everlasting oil, lit the place. Attendants were commissioned to serve there. The sanctuary became a pilgrimage site on fixed dates. Every year on the anniversary of Ay Aṭām Kujukrī they congregated there.[124] They prayed and offered the mummy expensive gifts and sacrificed animals. Hence all the kings descended from Ay Atām Kujukrī were named A.L.T.N [read *altun* or *altin*; *altın* in modern Turkish] Khan, which in Arabic means *malik al-dhahab* (the Golden King). They carried on in this manner for long years, for a period of thousands of years. During this epoch they lived lavishly and prosperously, not encountering enemies nor experiencing war or bloodshed." Jibrīl b. Bakhtishūʿ ends at this point.

b) al-ʿAynī: Genealogy and tribal division

The famous Mamlūk historian Badr al-Dīn Maḥmūd b. Aḥmad (1360–1451) acquired his noun of relation al-ʿAynī from the town of ʿAyntāb (Gaziantep in modern Turkey), his birthplace on the Turkish-Arabic frontier. Due to this he was fluent in Arabic and Turkish. Like other contemporary scholars, he had also some knowledge of Persian. Moving southward to the heart of the Mamlūk Sultanate, al-ʿAynī established fruitful contacts with religious, academic and governing circles. His linguistic skill smoothed his way to gaining the patronage of several high-ranking Turkic commanders and sultans.[125] In this position he composed the two books that contain the narrative of early Turkic history, which is translated here.[126]

All human beings are the progeny of Adam. Yet almost all his descendants vanished in the deluge, except those who boarded Noah's ark, who were rescued.[127] According to the general opinion of scholars and exegeses they numbered eighty souls. Noah's ark rested upon the Jūdī Mountain.[128] The passengers disembarked at the place of landing (*apobaterion*) and built a village there. This location is named the village of eighty and is situated in Upper Mesopotamia.

Noah lived after the Deluge for 350 years. He died at the age of 1780 years. All the passengers of the Ark passed away, except Noah and his three sons: Shem (Sām), Ham and Japheth and their wives. On his deathbed Noah nominated Shem, who was born 98 years prior to the Deluge, as his heir. Noah divided the Earth between his three sons. Shem received the central (or middle) lands which comprise: Arabia (Ḥijāz), Yemen, Syria and the Holy Land (*bayt al-maqdis*) and Egypt, and the great rivers: the Nile, Euphrates, Tigris, Oxus (Jayḥūn; Amu Darya) and the Jaxartes (Syr Darya; Sayḥūn). Ham was allocated the lands of the West (North Africa and Europe) and the lands west of the Nile (sub-Saharan Africa) unto the shores of the west wind (*manḥar riḥ al-dabūr*).[129] Japheth was allotted the South and the eastern lands. All the experts in genealogy agree that all the nations issued from these three sons and spread throughout the Earth. They also concur that Japheth was the eldest of the three, Ham was the youngest son and Shem the middle son.

As for Japheth, he had seven sons, so the Bible (*al-tawrah* i.e. Torah) says.[130] The eldest son of Japheth was Gomer, who fathered three: Riphath, who is the ancestor of the Franks. They inhabit a vast territory that encompasses 150 central towns and additional provinces. The second son is Ashkiyan (read Ashkenaz) who are the slaves (or Slavs; *al-Ṣaqāliba*). The third son is Togarmah (*tūgharmā*) who is the origin of the Turks. So at least maintains one opinion. Another son of Japheth was *Māghū'* [read Māghūgh, i.e. Magog], who is the origin of Yājūj and Mājūj (Gog and Magog), who are the Mongols. Tubal fathered the people of China. The correct position is that the Turks are the progeny of Gomer. Some say Turk was the son of Japheth.

Originally the Turks consisted of twenty tribes.[131] Each tribe encompasses numerous clans. The westernmost tribe, which was the closest to Byzantium, was the Bajnak (Pečenegs),[132] and then the Qipjaq or Qibjaq, Oghuz (young bull),[133] Ymalk, Bashghurt, Ysmal, Qay, Yabaqu, Tatār or Tatar or Ṭaṭar, Kirkiz, Jikil (Chigil), Tukhsi, Yaghma,[134] A'zaq/Aghraq, Jrq, Jaml, Ayghr, Tankat, Khitay or Khiṭya or Khiṭa who are called China (*ṣīn*), and Tughaj that are also named Masin.

The Turkoman clans, twenty-two in number, are a branch of the Oghuz tribe. Each tribe has its exclusive insignia and a distinctive emblem [is placed] on its animals and tools, which makes it possible to identify each one of the groups. The Qiniq (Kınık) is the principal Turkoman tribe. The kings and sultans are from this tribe. The offspring of Saljūq b. Duqāq belong to this tribe. Their descendants seized the Byzantine territories in Asia Minor. The Saljūqs crossed the frontier and penetrated the Abode of Islam in the month of Rabī' I, 465/November 1072.[135]

The insignias (*'alama* i.e. tamgha)[136] of the Turkomans (Turkmāniyya) tribes are: Qiniq (Kınık or Kïnïk)[137] the principal Turkomans ; Qaba' or Qaban ; Bāyandar (Bayındır; Bāyundur)[138] ; Uwā, some say Yuwā ; Salghur, some say Salur ; Afshār, some say Awshār ; Biktilí (Baktili), some say Bikdilī ; Bukdhur (Bukduz) ; Bayāt ; Yazghar (Yazghir), some say Yazar ; Aymur ; Qarā Ilik ; Ulqā Ilik (Alqā Buluk) ; Akdar or Yakdar (Ikdir) ; Urkar or Yurkar (Urakir) ; Tūtar ; 'Ūlā Yundulugh ; Tūkar or Dukar ; Bujnak (Bajanak) ; Jūlduz or Juvuldar ; Ḥabatī (Jabni) ; Jaruqlu' or Jaruqlū (Jaruqlugh) who number only few and whose brand is unknown. This is the list of twenty-two men. Each one of them is the ancestor of a clan.[139]

The origins of this organization is in an episode that happened while Alexander the Great (Dhū al-Qarnayn i.e. the man with the two horns)[140] advanced towards the country of the Turks. At that time the king of the Turks was a man named Shu (read: Shūl or Sūl) who commanded an enormous army. Yet, under the darkness of the night Dhū al-Qarnayn succeeded in surprising this huge camp that fled in great confusion to all directions. These twenty-two men were slow in reacting. Unable to flee, they remained in the encampment. Alexander recognized that they were intelligent men and said in Persian: "They are *Turk Manan*", which means those people resemble the Turks.[141] So the name Turkoman stuck to these people and they are known by it from that day till now. Nonetheless, to make the pronunciation smoother they omitted an "n" from the "Manan." The clans of the

Turks include the Khuttal, Usrūshana, Sughd, Khizlikh, Ṭoghuz-Ghuz, Ghuzz, Khirlikh, Mongol, Butayta, Toghuz-Ghuziyya [again!], Khirkhīz (Kirgiz), Barghaziyya, Kīmäk (Kihākiyys), Jighir (Jikil/Cigil), Jāmāt, Khalkh, Yarghāniya, Khazar, Mughān, Guraana, ʿAllān or Allān.

The offspring of Saljūq b. Duqāq are from the Qiniq. The first king amongst them was Ṭughril (Toghril) Beg b. Mikāʾīl b. Saljūq b. Duqāq. The first to cross the Jayḥūn River and to penetrate the Abode of Islam was Alp Arsalān b. Chaghrī Beg b. Dāwūd b. Saljūq b. Duqāq. He commanded an army of two hundred thousand riders. The Saljūq rulers of Asia Minor were his descendants. The Saljūqs crossed the frontier and penetrated the Abode of Islam in the month of Rabiʿ I 465/November 1072. The last Saljūq sultan to govern Anatolia was ʿIzz al-Dīn Kay-Kawūs b. Kay-Khusraw b. Qilij Arslān who died in 677/1278–79.

c) Ibn Ḥassūl: The superiority of the Turks over other regiments (*kitāb tafḍīl al-atrāk ʿalá sāʾir al-ajnād*)

Ibn Ḥassūl (d. 1058) served, as was already mentioned in the introduction, as a secretary of the early Saljūq dynasty.[142] A polymath, as attested by his contemporary the anthologist al-Thaʿālibī (961–1038),[143] he composed his *Superiority of the Turks over Other Regiments* for Ṭughril (Toghril) Beg as an attack on the Būyids and as a tool to legitimise his lord.[144]

<26> In the name of Allah the Merciful and Compassionate.

Praised be Allah the Lord of the Universe. God bless our Prophet Muḥammad and all his family, and grant them salvation.

O Lord, I crave to find in you refuge from tribulations (*fitna*)[145] of evil-doing and wrong-saying, from the degradation of error and prattle, from stumbling and mistake,[146] from fatal mischief caused by falsehood and mendacity, from the disastrous calamity of rage and grudge.[147] O God, preserve me from the viciousness of disputation[148] and hypocrisy, from falling into the trap of baseness and impudence.

O Lord, I crave to find in you refuge from catastrophe caused by the whim of (carnal) desires,[149] from the forces of covetousness and their sway.

O God, keep me away from allegations that I am not acquainted with and not aware of, from information that I do not understand surely and will never grasp even after assiduous study.

I crave to find in you a safeguard against sinking into false discourse[150] that the seeker of salvation survives[151] and from exposure to <27> prolonged torment resembling trial by ordeal. O God, keep me from drawing closer to [wrongdoers] who disgrace religion and morality (*dīn wa-muruwwa*),[152] and safeguard me from the maladies that strike the person who has committed reprehensible sins, who should be reproached and condemned as faulty.

O Lord, I beseech your protection from any fault,[153] keep me from doing wrong in any of my deeds, those new ones that I shall carry out and those that I have performed and shall repeat.[154] Grant me success in all my endeavours and all my deeds,[155] those that I shall build or shall ruin. Guide me so that everything I say and report is reliable and true.[156] Grant me loyalty (*amāna*)[157] in all my endeavours, both visible and covert,[158] so that everything I speak and utter is well directed (*tasdīd*),[159] so that my gaze does not convey oppression while my intention is equity (*inṣāf*),[160] and I will not use biased[161] scales though equal balances is my purpose.[162]

O Lord, keep me from being like those who sell their recompense in heaven in exchange for earthly benefits,[163] like one whose passions have led him astray,[164] like the man who seeks the company of slanderers, who speaks ill of others, who acts with injustice against the absentees, preferring to render a favourable verdict to those present, who upbraids people that the doom of demise (*maniyya*)[165] is closing in on them and who stick together in hope of being resurrected on the Day of Judgement.[166]

In God's original religion (*fiṭra*)[167] all those things are reprehensible[168] and are considered a stumbling block. Such behaviour is considered despicable baseness that leads to haughtiness and pleasing the verdict of transgression (*qadr*).[169] Moreover, these traits fit only a man whose religious belief is uncertain, who lacks judgement,[170] who is removed from the way of rectitude,[171] who persists in his wrong course, who exceeds his proper measure[172] and does not accept divine predestination (*qadar*).[173]

al-Ṣābī

You, may God prolong your life, mentioned the book named *The Crown* (*al-Tājī*)[174] <28> written by Ibrāhīm b. Hilāl al-Ṣābī,[175] saying how its beautiful language and wonderful writing excited you,[176] adding that you were pleased by the author's investigation of the history of the people of Daylam and their genealogy,[177] his knowledge of their rise and the circumstances that led to their successful establishment as the de facto rulers of the Caliphate (*dawla*).[178] In his book al-Ṣābī mentioned the Daylamites' offensives in the battles they initiated, the praiseworthy deeds they accomplished,[179] the determination that enabled them to realize their aspirations and the key positions (*maqālīd*)[180] that they seized. Leading their blessed multi-ethnic armies they secured their aims. [He reported on] the elevated battlegrounds (*maqāmāt*)[181] and the low valleys (*khuṣūm*)[182] where they advanced against their opponents, how they surmounted afflictions and conducted wars, how they confronted their adversaries and how they overcame the ravages of time. In his technique he, al-Ṣābī, reached the acme of elucidation that he aspired to accomplish and most elegantly addressed the topics that his writing revolved around.

In his presentation al-Ṣābī aimed to praise and extol the person named ʿAḍud al-Dawla ["the mighty hand of the [ʿAbbāsid] dynasty"], the son of Rukn al-Dawla ["the pillar of the [ʿAbbāsid] dynasty"].[183] He eulogized (*taqrīẓ*)[184] and praised him. He panegyrized ʿAḍud al-Dawla and credited him with deeds he was aware of, but also ascribed to him purported activity and praised him for that. He went so far as to claim for ʿAḍud al-Dawla the perfection (*kamāl*) of which no one can boast[185] except Allah, may He be praised and honoured.

Al-Ṣābī ascribed to ʿAḍud al-Dawla a wondrous ability to perform miracles (*iʿjāz*),[186] which the Lord granted uniquely to His prophets and messengers, rest their souls, especially to the Prophet Muḥammad, may Allah pray for him and the members of his family and grant them peace. Al-Ṣābī wrote with such exaggeration that he mixed truth with falsehood, veracity with lies, aggravation and leniency, compassion with harshness. He spoke in his book <29> at length and without hesitation, as one who thought himself secure[187] from opposing voices that would pass judgment on his word, as an author who seeks to evade critics that would dispute his views.

The Buwayhids

I swear on my life that al-Ṣābī should not be reproached[188] for the matters that he chose to include in his book, nor for his motives in writing it. This is so, because al-Ṣābī had suffered greatly at the hands of ʿAḍud al-Dawla and this made him fearful,[189] yet in contrast to his expectation he was surprised by the favour ʿAḍud al-Dawla showed towards him. ʿAḍud al-Dawla had borne rancour against al-Ṣābī because he was offended by a sentence that this author had used when he wrote in the name of the ʿAbbāsid Caliph al-Ṭāʾiʿ (r. 974–991)[190] to ʿAḍud al-Dawla, while the latter was in the province of Fars (Persia) considering whether to march on Baghdad, and to drive his cousin Bakhtiyār, nicknamed ʿIzz al-Dawla ["the might of the dynasty"], the son of Muʿizz al-Dawla ["the valour of the dynasty"], out of the capital city.[191] Bakhtiyār succeeded in consolidating his position as the master of orders and prohibitions, and ʿAḍud al-Dawla competed with him for influence at the majestic court of the Caliph.[192]

In addition ʿAḍud al-Dawla was affronted by the language of instruction used by al-Ṣābī and wished to punish al-Ṣābī on account of trifling matters that happened incidentally in the past, at the time of Rukn al-Dawla (947–77), and which caused tension between them. This included the differences that followed al-Ṣābī's giving ʿIzz al-Dawla Bakhtiyār the royal title "King of Kings" (*Shah-an-Shah*).[193] By elevating ʿIzz al-Dawla to the highest status, al-Ṣābī departed from the established protocol of the Buwayhid house, which decreed equality among the princes. This weighed heavily on ʿAḍud al-Dawla,[194] who loathed al-Ṣābī's coarseness. He resented him and conceived vehement hatred for him. ʿAḍud al-Dawla knew that al-Ṣābī had preferred these words,[195] and in writing the epistle chose to employ them in order to stir up dissension. Consequently, he was aware that al-Ṭāʾiʿ (d. 991) and ʿIzz al-Dawla had no part in the composition of this epistle, and that they were not responsible for the style al-Ṣābī had chosen nor for the painful and ignominious words he had used in his letter.

Subsequently ʿAḍud al-Dawla achieved his purpose, defeated ʿIzz al-Dawla and killed him [in May 978].[196] <30> Furthermore, by ordering his blood be shed,[197] ʿAḍud al-Dawla insulted ʿIzz al-Dawla Bakhtiyār's inviolability (*ʿiṣma*).[198] This distressed al-Ṣābī, who was pursued by terror, and being on the brink of death, he was eager to placate ʿAḍud al-Dawla. Standing on the edge of the abyss and hoping to rescue his soul from ʿAḍud al-Dawla's tusks and claws he wrote the book entitled "the Crown" (*al-Tājī*).

After al-Ṣābī felt secure from the awe of death's verdict and his reinstatement in the governmental services, he did not hesitate to employ the same old discourse once more. Thus, he added and reduced, lengthened and shortened, blamed and praised, stinted and expanded, attributed niggardliness and ascribed generosity, ascended and descended, extinguished and inflamed.

We should concede that it befitted ʿAḍud al-Dawla to refrain from false self-glorification. He abstained from fake boasting and did not acknowledge any virtue that was not a quality of his. He did not cull any fruit that he had not won and his intellectual capacities were not so weak that he could be deceived. Despite this, as regards to his genealogical tree he was deluded,[199] being misled like a wily horse whose reins are slack and whose harness wobbles. Although previously betrayed by al-Ṣābī, ʿAḍud al-Dawla expressed his fidelity to him. He shielded him carefully even though al-Ṣābī wickedly deviated from justice.[200] He conducted himself with forbearance (*ḥilm*)[201] even though al-Ṣābī's piercing criticism insulted him.[202] ʿAḍud al-Dawla guarded his honour, although al-Ṣābī stripped himself of modesty and his nakedness was exposed.[203]

With God's permission, I clarify this so that everyone may learn about my fair passing of sentence and the accuracy of my descriptions of the different subjects. Three I shall call upon to attest to the truth of what I shall say: unveiled true facts,[204] indubitable and solid guides that avoid confusion, and peers that share with me the knowledge of this branch of science.[205]

Daylam and Gilan

The transmitters of historical information and the narrators of chronological traditions know that the Daylamites and the Jilanites [Gilanites] are two groups whose lands of origin are known, and whose settlements <31> are easily recognised, as is their distinctive physiognomy.[206]

The Jilanites [Gilanites] are groups and tribes whose boughs split and whose knots branch off,[207] while the Daylamites are subdivided into two. One group is the Istaniyya [Justanids] and the second, the Lanjiyya. The Istaniyya live in inaccessible rocky slopes, fortresses and very steep mountains in the country of Daylam.[208] Since ancient times their leaders have belonged to the house of the Wahsudaniyya [Vahsudan],[209] and they still lead them to this day. As for the Lanjiyya, they inhabit the stony steppes and the valleys of the country of Daylam. In the past their leaders were the Justaniyya, <32> and this is the situation today as well.

The Daylamites were infidels until the noble descendants of the Prophet Muḥammad (*ashrāf*),[210] the progeny of ʿAlī (*ʿalawiyya*),[211] settled among them and succeeded in securing their support. They preached Islam among them, exhorting the Daylamites to believe in the religion of Allah and the tenets taught by Muḥammad, the Messenger of Allah, as well as to adhere to the leadership (*imāma*)[212] of ʿAlī b. Abī Ṭālib.[213] The *ʿalawiyya* missionaries taught them to prefer ʿAlī's exclusivity and his superiority over the entire Islamic congregation. No companion of the Prophet Muḥammad could be associated with him.[214] Because of this, most of the Daylamites are Shīʿīs. Only a handful among them are Sunnīs. These people adhered to the Sunnī school of Islam because Muslim savants from the Ḥanafī and Shāfiʿī legal schools[215] mingled among them. The Gilanites are mostly Sunnīs. This is so because their conversion to Islam took place earlier. The Daylamites are mostly Shīʿīs, as they were converted to Islam by the Nāṣiriyya-Zaydiyya.[216]

<33> They, the Daylamites, benefited from the teaching of both Sunnī as well as Shīʿī religious scholars. They taught them about the unique position that the offspring of God's Messenger hold among the Arabs and the special status that the Nāṣiriyya-Zaydiyya lineage enjoys in this family tree, their noble nature, origin, and rank.[217] They were instructed to enjoy boasting about the refuge they had given to this progeny, that they should express their pride in poems, declarations and propaganda whenever they prepared for battle and organised themselves to fight adversaries, who lacked this unique and noble distinction.[218]

The Genealogy of the Arabs

Yet, no person ever heard the Daylamites claiming that they belong to the family of the Prophet. This idea never occurred in their minds. They did not sing of it, neither in elegant poetry nor in satires. They never stated it, neither in their war slogans nor in sermons. The gifted among the Daylamites were not angered when they heard songs calumniating the Banū Ḍabba[219] and when the merits of this tribe were enumerated, they did not think this referred to them. They were not dispirited when abuse was hurled at the Banū Ḍabba.[220]

Never in the history of the Arabs, neither in their history before Islam nor during the recent Islamic period, has there ever been an account of anyone from the Banū Ḍabba tribe who moved to the territory of the Daylamites, settling in their encampments or in proximity to them.[221] A catastrophe never came upon the Banū Ḍabba forcing them to emigrate from their homeland in Iraq, Syria, or from their isolated territories in the remote deserts in the western regions.

The Banū Ḍabba tribe supported[222] Muʿāwiya b. Abī Sufyān,[223] whom they all joined in avenging the blood of ʿUthmān,[224] may God be pleased with him, and declared war against ʿAlī b. Abī Ṭālib, peace be upon him. The proof of this lies in the stanzas of the poem of one of their spokesmen, who declaimed:

> "We the Banū Ḍabba participated in the battle of the Camel /[225]
> With the sharp edge of our spears we mourned the death of ʿUthmān Ibn ʿAffān. /
> Give us back our leader, and then he would be honoured with respect."

However, al-Ṣābī did not hesitate to characterize this ethnos with a false lineage and freely provided the Daylamites with an Arab ancestor. He was able to trace this refuted noble lineage <34> because the genealogical tree of the Banū Ḍabba is utterly frail and untenable. Their number is great,[226] they are neither like the Arab tribes such as the Quraysh, Tamīm, Ṭāy' and Qays, nor similar to Khindif and 'Uqayl,[227] which are the central Arab tribes. They are not counted among those tribes that occupy an elevated noble place in the Arab genealogical tree.

If al-Ṣābī could have succeeded in associating the Banū Ḍabba with the lineage of the Quraysh tribe, he would have done so and subsequently 'Aḍud al-Dawla would have been given the qualifications to fit the leader of the Muslims perfectly and would have been established as the legitimate head of the Islamic Caliphate.[228] By this means 'Aḍud al-Dawla would be enabled to array himself with this patrician capacity even though he lacked it. Al-Ṣābī was prepared to toy with these ideas, even though 'Aḍud al-Dawla did not boast of that grandeur.

The first grave mistake that al-Ṣābī made in his book *al-Tāji* was tracing the genealogy of this ruler (*wālī*)[229] back to a false and dubious lineage. Al-Ṣābī put 'Aḍud al-Dawla in the position of a person who is deceiving himself and is dazed in perplexity.[230] Afterwards he acted as if he had ignored the Arab genealogical tree. He removed 'Aḍud al-Dawla from it and positioned him within the Persian genealogical tree of Bahrām Jūr.[231]

It is known with certainty that the Arab and Persian genealogical trees are well separated. Experts in genealogy report that these two branches join only in Abraham the Patriarch. It is said that the Persians are the descendants of Isaac, while the Messenger of Allah is the progeny of Ishmael. This is the source of the statement by the Prophet Muḥammad: "I am the son of two to be sacrificed."[232] By this he meant Ishmael[233] and his father, 'Abd Allāh.[234] 'Abd al-Muṭallib, the grandfather [of the Prophet Muḥammad] wanted to sacrifice his son ['Abd Allāh] until he was redeemed by a payment of many she-camels that saved his life. <35> [This happened] after the divinatory arrows of the wheel of fortune (*qidāḥ*)[235] turned, and the arrow rotated between slaughtering or saving ['Abd Allāh's] life, between sacrificing him or redeeming him.[236]

If, as al-Ṣābī argues, the Daylamites are Arabs, they must be from Ishmael's progeny. Therefore we need to know which forefather he [al-Ṣābī] is referring to, with whom the Daylamites branched away from the genealogical tree of Allah's messenger, the lineage that al-Ṣābī claimed that 'Aḍud al-Dawla belonged to, the privileged ancestry with which he honoured him. It is well established that the nearer a person is to the Prophet Muḥammad, the greater he is in nobility, his lineage is clearer, his fame is more conspicuous and his roots are more aristocratic.

The genealogical tree of the Quraysh indicates that the Prophet is related to all the branches of the tribe via al-Naḍr b. Kināna. Therefore every one who is a member of Quraysh is suited to leadership of the Islamic polity (*imāma*), as the saying of the Prophet [in the *ḥadith*] goes: "The legitimate leaders of the Islamic community are exclusively from the Quraysh tribe."[237] As regards the clans of Banū 'Abd Manāf, Hāshim and 'Abd al-Muṭallib,[238] they are the closest kinsmen [of Muḥammad], on both the paternal and the maternal sides.[239] This is exactly what the Messenger of God used to boast of, the honours he enjoyed through his ancestors when he said: "I am the son of the women named ʿAwātik[240] and the son of the women named Fawāṭim."[241] ʿĀtika was the daughter of ʿAbd al-Muṭallib, and Fāṭima was the daughter of ʿAbd Manāf.[242]

I do not know when the Daylamites boasted of this blood-tie or when they adhered to the idea of this most solid kinship.[243] Has anybody heard the name of Daylam mentioned in ancient poetry except in the poem (*qaṣīda*) of ʿAntara [Ibn Shaddād] al-ʿAbsī,[244] when he said: "water rivulets gushing out from the Daylam water-pool."[245] <36>

This is because when Fīrūz[246] fought the Arabs and raided their camps, he brought with him the Daylamite army in the same manner that scabby ravens flock in a gathering place – some are enduring and others ailing. They [the Daylamites] trembled [because of the Arabs'] ambushes and surprise nightly attacks.

The Legacy of Iran

Experts in genealogy,[247] learned scholars and historians of antiquity,[248] agree that after the period of Afrīdun,[249] the [Persian] kingship split into branches.[250] He is the one who killed al-Ḍaḥḥāk and conquered his kingdom. Al-Ḍaḥḥāk [Dahhak; Zohak; Zahhak][251] is the one named in Persian Yūrāsaf [Aperasaf; Buhrasf].[252] The Persians narrate about him imaginary tales, which they believe, and foolish fables, which they cling to. Such is the legend of the two snakes nestling on his shoulders. They are supposed to have wrapped themselves round him. When these two serpents became hungry, they became merciless, stinging and pricking. They fed only on the brains of human beings. Because of this bane that struck him, he [Yūrāsaf] is said to have killed two young boys from his community each day.

Logical men would reject these false stories. Accepting them as truth is shameful and should be considered improper. Indeed this man was a felon, a violent robber, an unruly tyrant and obstinate. He had a notoriously bad reputation, a dreadful character from whom the public only wanted to escape and his enemies to surmount him.

It is well established that the historical narrative of the early period that preceded al-Ḍaḥḥāk's time is an accumulation of dubious and ambiguous stories, a collection of wild and uncontrolled fables.

The well-founded historiography starts, without a doubt, only with the period of the time of Afrīdūn, <37> the conqueror of the world.[253] The empire of this mighty king encompassed vast zones in the East and the West. He ruled over seas and lands. Afrīdūn had three sons: Tūr, Salm and Īraj. He divided the land over which he ruled between them and gave a portion to each of his sons.[254] Salm was allocated the West, the land of the Greeks (*rūm*) and the area beyond it up to the territory of the Arabs. Īraj received the centre of the fourth clime, which is the best region and the most praiseworthy.[255] To describe its qualities would exceed the limits of our purpose and for this reason we skip it in the present book. Tūr's area was the East and also the northern countries up to the border where civilisation ends.[256]

Yet for a long period Salm and Tūr conspired against Īraj and tried to ruin him.[257] They overpowered and killed him. Then Manūjhar (Manushihr; Manchoor) emerged and claimed vengeance for Īraj's spilled blood. He found remedy by crushing his opponents. He took by force the country that was meant to be his inheritance. He conquered and appropriated it. Deep loathing broke out between the descendants of Tūr and the progeny of Īraj, and indeed the animosity between them still exists to this day. The two clans cling to the hostility that they harbour and keep it alive, generation after generation. If someone should forget about it, they would remind him of it and renew the flames.

This bitter enmity has even intensified ever since the king Frāsiyāb (Afrāsiyāb; Afrasiab) [of Tūrān] travelled to these areas, subdued their people and abused them, extorting taxes from them and exploiting their lands.[258] For twenty years he stayed and ruled over them, until Kay-Khusrau (Khosrow) and Rustam (Rostam) al-Zāmilī,[259] his two indefatigable opponents, stood up against his tyranny and pulled the rug from under his feet and toppled him. The animosity between the two races has continued from that time to this.[260] Fighting followed day after day, and one war succeeded another. These matters were explained in Arab and Persian writings, both in prose and in verse.

<38> The overall objective of my work is to provide evidence as regards the genealogical tree of the Turks, and to prove that they are the descendants of an illustrious and conquering ruler. Dangerous and impregnable castles that are visible in the Realm of Iran (*Īran-Shahr*)[261] have struck fear in his heart and in the hearts of his descendants. They were not seized until centuries later you [Ṭughril Beg] took possession of them.

Among the Hellenic-Greek descendants of Salm was Alexander the Great, who is also nicknamed the Two-Horned (*dhū al-qarnayn*).[262] He stormed the Iranian homeland and in a *tour de force* marched around it. For almost two hundred years the Hellenic Greeks ruled the Iranian plateau. They governed according to the guidelines and principles of government outlined by Aristotle,[263] Alexander's vizier (*wazīr*).[264] However, it is an intellectual error, as well as a religious sin, to shed innocent blood. Belief forbids illicit killing and after deliberations, reason (*ʿaql*)[265] reaches similar conclusions. Killing a human being is allowed only as vengeance, in retaliation for a murdered soul.

Yet, this land of Iran is of the quality that stood firmly against these sorts of people.[266] How did he [Aristotle] imagine he could mow a cultivated field whose plants grow up quickly and within a short time is covered by replacements?[267] But, regardless of this fact, he had conceived an erroneous idea. According to this, the land of Iran should be divided between them [i.e. the Hellenic Greeks] and they would all be of equal status, so that they would be jealous of each other and compete and fight against each other so that the vigorous would win.[268] This would secure the leadership and the kingship in the hands of Alexander, and those of his progeny and kin who would inherit him.[269] By these means the Hellenic Greeks ruled the Iranian homeland for a little over two hundred years. These sovereigns of Iran were called the petty kings (*Diadochian*).[270] Their rule lasted until Ardashīr b. (Bābāk) Papak emerged.[271] He cancelled this principle and annulled this stratagem.[272]

<39> This Ardashīr seized the kingship thanks to his courage, heroism, forbearance, common sense, intelligence, wisdom, leadership and statesmanship. He has no blood ties with any king from the past worth mentioning and is not related to any leader who possesses servants. For this reason Ardashir announced: "My genealogy starts with me while the dynasty of my opponent ends with me." He made laws to prohibit flaunting one's lineage, blood ties and maternal ancestry if the person did not also have honourable traits and civilised manners.

Ardashir was the first king of the Sasanian dynasty and his progeny after him ruled over a vast kingdom until Islam emerged and obliterated their traces,[273] took away their kingdoms[274] and condemned them to misery, except for those who came under the aegis of Islam and its protection. He, Ardashir, was the first to rule by righteousness and to govern by justice. He was agreeable to his subjects and eased the burden of taxes on the land. One of his wise decisions was to sign a truce (*hudna*)[275] with the kings of the Turks. He withdrew from large sections on the borders of his kingdom and gave these territories to them. Ardashir also instructed his successors to continue with fair conduct towards the Turks.

The virtues of the Turks

Now we shall address ourselves to a review of the attributes and qualities of this people, by which I mean, the Turks: their nature, character, qualities, customs, methods and ways. We are not according them anything of which they are not worthy and we will not strip them of what it is their right to be proud of. We are neither adopting the approach of ethnic solidarity (*taʿaṣṣub*), nor writing against them.[276] We are neither seeking advantage by reporting favourably on them, nor looking in vain for their benevolence. Rather, our book will be exposed to reviewers who <40> will examine its contents and statements meticulously.[277] Moreover, they are serious commentators who will analyze and observe critically. We ask God to protect us from hypocrisy, which is condemned by all the laws and beliefs, and outlawed at all periods and times.[278]

We shall start our description by mentioning their heroism. Their bravery is among the noblest traits of the rational mind (*al-nafs al-nāṭiqa*)[279] especially when it is directed towards the protection of the [revered] women and the fight against foe and enemy. They are particularly eager to face adversaries, to use arms with the purpose of achieving absolute victory and avoiding abuse. Protecting the tribe and the family, the Turks are committed to a worthy zealousness and honourable modesty. Thanks to this the [Caliphate's] provinces are well organised and the roads are safeguarded.[280] No other people demonstrate sturdiness similar to the endurance displayed by the Turks and no other people has shown comparable persistence and readiness to go to great lengths to achieve their objectives.[281]

God has created them in the shape of lions: with wide faces and flat noses. Their forearms are colossal and muscular and their characters grim.[282] Only seldom is it possible to see among them negative traits such as slim hips, smooth hairless cheeks and outsized eyes. Seldom would they seize food by force and rarely would they refrain from accomplishing obligations. Unlike other army units, in this ethnic group it is enough to discipline one of them in order to hold the others in check. Shouting directed against a particular person among them will not drive away the entire group and restraining one of them will not affect the others.

They consume only meat and do not wish for any substitute, even if it is dripping blood or filthy. This does not affect them and they are not deterred from eating it. They do not consider meat worthy of consumption unless they have used force to obtain it,[283] and do not enjoy prey if they were not fatigued by a wild chase. <41> This is the habitual behaviour of predatory beasts of prey, lions that pursue their prey.

They are used to the wilderness and to the dry steppes. In situations of food shortage[284] and difficulty their endurance arouses wonder. Even during the most violent confrontation or in the face of overwhelming hazards they behave as someone who is enjoying his leisure, as someone taking pleasure in pursuit of a frightened deer or a wild ass that has no owner. It is as if the thought of exhaustion does not reach them and does not occur to their minds. They continue with their energetic activities galloping their horses, climbing high mountains, riding in the face of dangers, making forays into narrow canyons and penetrating deep into unknown countries.[285]

Islam made it a religious duty to raid (*ghazw*= razzia) the infidels among the Turks,[286] and decreed likewise regarding other heathen groups and folks in different lands.[287] This attests to their noble qualities and their enormous importance. Those among them who were taken prisoner wanted only one thing: that their new master should treat them with equality as regards food and drink, dress and riding.[288] The Turkish prisoner would not be accustomed to humiliation by his master, as is the case with other enslaved people (*riqq*). He would resent his master employing him in servile labour or ignoble tasks in which his master engaged others who as prisoners have been carried from distant lands to the Abode of Islam (*majlūb*),[289] and who are employed in servile duties such as sweeping houses, looking after the cattle and other wretched tasks assigned to slaves (*raqīq*) inflicted with an infidel's misfortune and rendered obedient by the oppression that has pursued them. We have never seen a Hindu, a Byzantine, an Armenian or anyone from any other enslaved people who remained unaffected; the limited ambitions of these bonded persons are well-known, as can be recognized by their restrained moves.

<42> When a Turkish prisoner of war is released from his bonds he will not be satisfied with less than being assigned the leadership of an army or the position of chamberlain or the head of a platoon or the command of a division. And this conduct is not limited specifically to Turks in Khurāsān, who dwell in this province that borders the Eurasian steppes and is close to the Turks' homeland. This is the situation everywhere; if you wish to have a distant example, take Egypt for instance,[290] although this country is in a region very remote from their birthplace and its people know the least about the civilisation and language of the Turks. Or for another example, take Iraq, which attracts soldiers from various ethnic groups (*aṣnāf al-ajnād*):[291] Arabs, Daylamites and Kurds, and was ruled by a usurper from Daylam who asserted his position by claiming for himself an ancient legacy and venerable merits.

Generation after generation, the ʿAbbāsid caliphs and Daylamite kings looked proudly towards them [the Turks], tried to overpower them, to act recklessly towards them, to challenge their rights, or cause them to subdue others. However, they were not blessed with the wisdom to set themselves apart from the Turks. They were not endowed with the strength to subdue the Turks, to suppress them or the might [to overpower them].[292]

The saying of God's Messenger: "Leave the Turks alone as long as they leave you alone"[293] is the best proof of their glorious ability, their evident strength, their fierce pride and their united stand.[294] Furthermore, the Prophet Muḥammad, God bless him and grant him salvation, said: "I am the messenger to the blacks and the reds",[295] in other words, to the Arabs and the Persians. Moreover he commanded: "Wage war against all the peoples until the word of Islam wins and all of them join the community of the believers."[296]

<43> The most extraordinary virtue of the Turks and their unique merit is that as a rule nobody ever saw a pure Turk who was afflicted by effeminacy (*takhnīth*).[297] This in spite of the fact that this shameful disgrace and abominable deficiency is quite widely spread.[298] We see it among all the peoples, especially among the Jilanites.[299] If, however, we find among the Turks a person that demonstrates femaleness (*taʾnīth*),[300] somebody who has espoused the ways of women in his manner, in his speech, in his hints, his dress or his adornment with jewellery, this man must be belong to those Turks who were poisoned by the disgraceful [conduct] of those neighbours amidst whom they dwelled in these remote Iranian provinces.[301]

The coming of the steppe peoples

Having pointed to some of the virtues of the Turks and their high merits that none can oppose or dispute,[302] let us go back and present the kings of past ages who were of a Turkic race or from countries that border on the Turks' territories in Central Asia. Such are the Samanid (Āl Sāmān)[303] kings of Khurāsān and others. Also, we would point to Subuktakīn [Sebük-Tegin the Ghaznawid, 977–997], his sons and grandchildren: Muḥammad, Maḥmūd and Masʿūd, may God be pleased with them all. How firmly they settled in their territories and courageously, with their own hands and strength, overcame their opponents and those who competed against them! How they with their own hands pursued their aims and achieved them! They illuminated their days and raised their banners.

This splendour continues still in these days when the territories of these rulers are joined together under the reign of our master, the Sultan of the universe, the King of Islam,[304] the most excellent (*ajall*)[305] and mighty Emperor (*shah-an-shah*),[306] the upholder of the faith, saviour of the Muslims, the radiant light of God's religion, the ruler of Allah's countries, the redeemer (*mughīth*) of God's servants,[307] Ṭughril (Ṭöghril, Toghrïl)[308] <44> Beg Abū Ṭālib Muḥammad b. Mīkāʾīl,[309] the supporting right hand (*yamīn*)[310] of the Caliph (*khalīfat Allah*),[311] the commander of the faithful (*amīr al-muʾminīn*). May God aid him and help him to win victories, strengthen his sovereignty, humiliate his enemies, and fix his banners firmly. He proceeded till his justice engulfed the Land and his reputation extended to all parts of Earth, from the East to the West.[312] He was given a sovereignty that none of those who will succeed him will achieve,[313] and he is clothed in splendour such as no king before him could claim. Every Muslim should obey him and stand beneath the wings of his authority. Both secretly and openly, a Muslim should be devoted to him and be loyal to his sovereignty. Day and night he should stand, bless him and invoke God to augment His support of Ṭughril, to sustain his fluttering banners and to display his emblems.

We shall mention a few of the things we saw with our own eyes and accomplishments of which we were told. We shall report on his achievements in implementing justice and manifesting benevolence.[314] He sought to please God by implementing equality and righteous conduct,[315] by being merciful towards his subjects, in consent as well as in wrath. Great was his resentment of kings, among his predecessors, who had applied themselves to amassing capital and property. We will explain this clearly, not blend our words with flattery nor taint them by greed. Lies have not been mixed in with our words. They will not contain hypocritical elements nor shall we use deceitful words to defame the reputation of others.[316]

God forbid, neither fear nor greed motivated us to adopt such an attitude. Neither apprehension nor hypocrisy nor flattery brought us to write this, nor the desire to please him. On the contrary, we were driven solely by the quest for integrity that he deserves, by devotion to justice for its own sake, striving for the truth as an objective in itself. <45> Only the wish to do justice to his genuine generosity, gratitude for the safety we felt during his reign, and the tranquillity we experienced under the canopy of his government.[317]

ʿAmīd al-Mulk al-Kundurī

The strongest motivation in recording this book, the warmest encouragement to compose it in a clear manner and to embellish it, derived from the fact that I relied upon the glorious leader, the chief (*sayyid*),[318] the helper (*muʾayyad*),[319] ʿAmīd al-Mulk[320] ʿImād al-Dīn Abū Naṣr Manṣūr b. Muḥammad [al-Kundurī], may God sustain his eminence. His [highly esteemed] commentaries on the contents of the epistle and its style, ascertained the truth of its aims and their intended meaning and objectives. He clarified the flaws he found in the epistle and refrained from confirming what I might have claimed in favour of high-ranking personalities. He subsequently translated the epistle into Turkish for the Sultan [Ṭughril Beg]. His [al-Kundurī's] impressive presence is one of the most gladdening facts of [Ṭughril Beg's] era. [Al-Kundurī's] presence enlightened this period and his great deeds enhanced the affairs of state. With his own hands (i.e. writings) and with his tongue he polished the conditions of our era. Virtues are portioned out among people and each receives his matchless feature (*ḥaẓẓ*), which is different from the fortune of the others. Wise men will bear witness that completion is scarce and that perfection is a miracle.

Science is divided between those disciplines that will save the Muslim in the Hereafter, and those that are needed for the existence in this world and the everlasting remembrance of God.[321] However, the only provisions necessary for man when he turns to God's mercy and His forgiveness, is the knowledge that God is one, so that he would know Allah truly and not confer on God attributes except those that increase His glory.[322] The only qualities that should be added are those attached to God's name, <46> qualities that strengthen God by exalting Him, glorifying His name and elevating it,[323] and presenting proof of His existence and His being one and unique, in such a manner that the evidence cannot be contradicted or disputed.

Next, a man must acquire knowledge of the prophetic missions of God's apostles. He must know that each of them manifested his divine mission by performing miracles, especially the Prophet who sealed the revelation of the Divine Heavenly Law[324] and whose prophecy abrogated all previous religions, namely, the Prophet Muḥammad, may God pray for him.

In addition to this, man must acquire the skill of deducing legal conclusions from the sacred preaching of Muḥammad. He must know that his religious practices (*sunna*) are superior to all religious traditions. A Muslim has to master the proofs that support the essence of the Sacred Book that the Prophet Muḥammad delivered to mankind,[325] a peerless and inimitable miraculous (*muʿjiza*) Qurʾān[326] that descended from God in heaven to Muḥammad's heart.[327] A Muslim must study the meaning of the rare words of the Qurʾān and its accurate reading, including the inflections, what the Book permits and what it forbids, its clear and ambiguous verses, its abrogating and abrogated verses. He also must master the reasons that these verses were revealed, whether implicit or explicit, hidden or evident, what is obligatory dictated by a direct written word or what is only a common practice (*sunna*).

Afterwards, a man must direct his attention to the study of the rules of grammar relating to explication and writing, to phrases in verse and in prose, to condensed discourse and prolix style (*iṭnāb*),[328] to allusion and rambling speech. All should be accomplished according to the rules of the Arabic language, which is the most noble among languages, and the people that speak it are superior to all peoples.[329] All these are issues pertaining to the sciences of religious tradition and belief.

In addition to these branches of knowledge there are related sciences that investigate the true essence of things created by God. Such are the knowledge of the celestial bodies[330] and the stars, natural acts and physics (nature), things that are adverse and those that are useful. For each one of these categories (*jins*= genus) of science there are book compilations and classified rules, and scholars who present a model to be followed, who serve as source of knowledge and to whom <47> [people] address themselves.

Seldom can one find a person who might hold an impressive position in, and leave a significant imprint on, one of these categories of sciences. If such a man had been found he would have been the remarkable leader to whom all eyes would have turned, the illustrious learned man, the supreme savant admired by all, the sharp and unlimited intellect. However, we have not heard of such a man, whose learning encompasses all these sciences, who is truly erudite in all the areas of knowledge, and who stands out among the common people or the elite, who has mastered the most difficult and simple subjects. We have never heard about any such person nor have we encountered anyone like this. We have never been told about such an outstanding figure, nor narrated about him.

However, here he is this illustrious leader, the head and chief (*ʿamīd*), ʿAmīd al-Mulk, may God protect his government and safeguard his tenure in office. He is the most expert in each one of these sciences and master of each, as if he were focused only on learning it and not involved in any other outside of it, as if during his life he did not spend time on anything else. He has achieved this despite his young years and manifold occupations, conducting the affairs of the West and East and ruling over the Arabs and the Persians, consuming all his time. This exceptional [achievement] would not have been possible had not God evidently granted him success. His accomplishments were not simply the consequence of fate that elevated and assisted. Rarely is a man endowed with such good fortune and the manifestation of a noble lineage that has no peer.

I would not be saying this unless I had conducted investigations, research, in-depth studies and inquiries, and had witnessed all the specific areas of these sciences, either in figurative expressions or by sound exploration, or an ample fortune was set up as an example. And indeed, the Prophet has truly spoken when he said: "When God wishes to favour a ruler, he assigns to him a loyal assistant (*wazīr*).[331] If he [the ruler] should forget something, he [the vizier] would remind him of it, and if he [the ruler] remembered, he [the vizier] would assist him in executing it."[332]

The confirmation of what I am saying is the known fact of the lengthy period during which he [Ṭughril Beg], may God extend his banners, held control of the imperial affairs, <48> and his fine legacy. Further evidence of his capabilities is his balanced judgement and sound attitude on any matter you may wish, be it the sound attitude to criminals or his generosity to people begging favours, be it the respect he showed to state officials or the grace he conferred on high-ranking families, or be it the distance he held from suspicious deeds. All these happened during the days he was lucky to benefit from the services of this remarkable and auspicious leader and supportive master concerned with the matters of his court; due to this his orders were executed throughout the areas that his government controlled. All paid attention to his [al-Kundurī's] advice.

One of the examples that proves this is what is narrated concerning the case of the chief and army commander (*al-amīr al-isfasālār*),[333] Sayf al-Dawla ["the sword of the dynasty"] Abū Isḥāq Ibrāhīm b. Yūsuf,[334] who was brought before him [i.e. the sultan Ṭughril Beg] in fetters. He stood before him, trembling in his bones, and fear took hold upon him.[335] Everyone expected that the sultan Ṭughril Beg, may God strengthen his supporters, would spill his blood and get rid of him by killing him, carrying out his revenge. But he rose and approached him, embraced him and accepted his request for forgiveness, whether its expression was sincere or deceptive. This he did in sharp contrast to the way in which ʿAḍud al-Dawla treated his cousin, ʿIzz al-Dawla Bakhtiyār, after he defeated him. ʿAḍud al-Dawla did not pardon him but killed him. Acting on an immoderate impulse for revenge he ignored his nephew's appeal and refused his plea, as if only his beheading, and afterwards presenting the head in a bronze bowl, would satisfy his craving for revenge. Where is al-Ṣābī now, what would he say concerning this magnificent truth and this noble pardon?

Even more outstanding than this incident is the next story concerning what he [Ṭughril Beg], may God preserve his sublimity, did in the case of the people of Isfahan. They had mutinied, displaying open animosity and waging war against him, <49> showing defiance and disloyalty.[336] Acting treacherously[337] and by means of provocative tongues, they abased themselves, causing him to move against them. Laying siege to the city and repairing the damage caused by the rebellion and the transgressions of the city dwellers amounted to huge financial costs. By rejecting an agreement with him they compelled him to conquer the city by brute force. He wholly defeated the citizens. Nevertheless, he did not show anything but clemency, which he granted them and which was so absolute that it embarrassed all the people and shamed the kings, his predecessors.

Even if this positive quality in the Sultan was the result of innate laudable eminence, of noble auspices, of a spirit that people strive to attain and to emulate, nevertheless it was the illustrious leader, the head and chief, ʿAmīd al-Mulk al-Kundurī, may God protect his tenure in office, who caused Ṭughril Beg to conduct himself with consideration and restraint. He [ʿAmīd al-Mulk] refined the sultan's manners by narrating to him anecdotes about the sages of the past, he cautioned him by telling him estimable moral fables behind closed doors, and thus he strengthened Ṭughril Beg's determination. This is in contrast to what ʿAḍud al-Dawla did to the people of Baghdad after he defeated them and killed Bakhtiyār. He did not spare the people of the town. As a punishment he allowed his soldiers to go on the rampage and to take revenge on the civilian population, he abolished their privileges. Motivated by vengefulness, the soldiers burnt the city. And this [took place] though the ʿAbbāsid court was within the city and the caliph was residing in Baghdad.

The genealogy of the Saljūq clan

And as regards the lineage of this Sultan [i.e. Ṭughril Beg], may God strengthen his support, if I must exalt his honour suffice it to say that his origins are not an anonymous slave owned by somebody or an obscure and unknown man, as was the case with some of his predecessors. We count among his ancestors Sarjuk [*sic*][338] who with his sword slashed the Khazars' king,[339] and with the mace he held in his hand bludgeoned him forcefully. He beat him so fiercely that the king's horse collapsed and its rider fell to ground, bruising his face, and lay inert. No one would achieve such an outcome unless he had a free and unfettered soul whose ambitions soar into the sky high above the bright Capella star. From him [i.e. Saljūq] their dominion started < 50> and their cause [i.e. dynasty] emerged.

And so due to the piety of this Sultan, may God give him strength, which I saw, and the accomplishments I observed, is his stable and long-lasting reign, its foundations were not shaken and its supports did not decline. I was present in his sublime court when I laboured to compose the edict of a nomination letter handed to the commander Zurayr b. ʿAlāʾ al-Dawla when he was appointed Governor of Jurbādhaqān.[340] The sultan imposed on him, among other stipulations, that he should keep his authority firm, that the governor's hand should not weaken, nor his entourage disregard him. The sultan, may God protect the expanses of his realm, commanded that the emir Zurayr should not rob the district's flocks or appropriate its people's private property. Ṭughril Beg added that if the governor did not act accordingly, he would not be pleased with him. All those present, judges, notaries, ministers and clerks were astonished because they had never witnessed such an exceptional and fine act as this and could not recall such eminence and admirable generosity.

The Almighty knows that from the very moment I had the good fortune to go into Ṭughril Beg's service and join those serving in his sovereignty, I have not seen him fail, either by disappointing a hopeful man or by speaking harshly to drive away a religious scholar. I have not seen him condone injustice nor sanction a crime. What should one think of a king whose simplest endeavours and lowest merits are such as these?

Following the detailed description of Ṭughril Beg's legacy and his conquests, accomplished with God's help, records of events arranged according to annals [will ensue], reporting on my experiences year after year from the time I, to my happiness, entered his service and obtained my share in his honourable deeds and his merciful contributions. Along with these, I will restore [honour] for ever to the most excellent ʿAmīd al-Mulk,[341] may God strike down his enemies, and will describe the deeds he accomplished while <51> I was far from his sublime gates.

Success derives from God and we trust in Him, our destiny leads to Him. With his judgement we are satisfied. His mercy will shield us.[342] May God bless the Prophet Muḥammad and his family and grant them peace and salvation.

God made it easy for me to finish this copy. The copying of the epistle was completed in mid-Dhū al-Qaʿda 649 of the lunar months [Sunday, January 28, 1252 CE].

3 THE TRADITIONAL MEDIEVAL HISTORICAL PARADIGM

a) al-Makīn Jirjis Ibn al-ʿAmīd: The beginning of the Saljūq dynasty

Jirjis Ibn al-ʿAmīd was an Arab-Coptic historian from Egypt who wrote a world history (*al-majmūʿ al-mubārak*) from the creation to the emergence of the Mamlūk Sultanate in Egypt and Syria in 1260. The history is divided into two sections: a) pre-Islam to the days of the Byzantine emperor Heraclius; b) Islam, from Muḥammad to Baybars.[343] The section on the Saljūq Sultanate reflects the contemporary rulers' interest in the past of a dynasty that employed their predecessors. The text has been known in Europe since the publication of T. Erpenius' edition (Leiden, 1625). The present translation is based upon a manuscript preserved in Istanbul and not consulted by Erpenius.[344]

He (the historian) says:[345] In the year 430 the dynasty of the Saljūq sultans began. [Marginal note: The roots of the Saljūqs are among the Turkic tribes of Transoxania in a location named Jund Bukhara, which is within the boundaries of the district of Bukhara. Their leader was Mikā'īl (Mikhā'il; i.e. Michael) b. Saljūq. He commanded a considerable number of Turks who did not obey anyone. His prominence had spread and the [Ghaznawid] sultan Yamīn al-Dawla Maḥmūd learned about Mikhā'īl's bravery and invited him with his clan to settle in Khurāsān. Later on he regretted it.]

Abū Ṭālib Ṭughril Beg Muḥammad was the first among the Saljūqs to obtain public recognition as a legitimate sovereign, his brothers – Cheghri Beg Da'ūd, Fayrūz, Arslān and Bayghū – were also acknowledged with him; they all were the sons of Mikhā'īl b. Saljūq b. Duqāq.[346]

Duqāq was a Turk, a brave man, a man blessed with good sense and a man gifted with organisational skill. He was the first of his tribe to convert to Islam. At that time Bayghū was the ruler of the Turks, and he used to consult with Duqāq and to listen carefully to his advice. He enjoyed his collaboration and counsel and regularly invited Duqāq to join him in his campaigns against the Turks.

After Duqāq's death his son Saljūq inherited his place. The king of the Turks gave him command of his armies. Yet, after a while and fearing that Saljūq would be a source of calamity, the rulers decided to kill him. Saljūq had learned about the developing danger and took refuge in the court of Shihāb al-Da'wa Hārūn b. Ilek [Khān], the ruler of the (Qarakhanid) Khanate (d. 992).[347] Saljūq asked Hārūn to support him in waging war against the infidel Turks and in raiding their territories. Hārūn provided Saljūq with a large army. However, during one of the raids an infidel Turk killed Saljūq, who died in the holy war as a martyr, aged one hundred and seven years. His son Mikā'īl (Mikhail) inherited his position. We already mentioned above the story of Mikā'īl and his sons. They operated in Transoxania, where a great number of Turks followed Mikā'īl. They obeyed only him and no one else.

It happened that the [Ghaznawid] sultan Yamīn al-Dawla[348] Maḥmūd b. Sabuktakīn (Sabuktegin), the emperor of India, crossed the Jayḥūn (Oxus; Amu Darya) River[349] with the intention of supporting Qādir Khan, the ruler of Transoxania. Mikā'īl b. Saljūq met with sultan Yamīn al-Dawla Maḥmūd during this occasion. The sultan was impressed by Mikā'īl's zeal and courage, as well as by his authority over his tribe. Mikā'īl asked the sultan's permission to accompany him and requested his authorization to reside in Khurāsān, on the grounds that after the withdrawal of the sultan he, Mikā'īl, would guard the territory from approaching enemies. Yamīn al-Dawla Maḥmūd refused and Mikā'īl reacted angrily. This roused the sultan's fury and he ordered that Mikā'īl be arrested. Following this incident, the sultan Yamīn al-Dawla Maḥmūd returned from Transoxania to his headquarters. His army carried the prisoner Mikā'īl in a cage. Mikā'īl's clan and companions escorted the withdrawing army and eventually settled in the valleys of Khurāsān.

Regretting that he had permitted the Saljūqs to remain in Khurāsān, the sultan Yamīn al-Dawla Maḥmūd passed away after a short period. He suspected their intentions and was worried by the positions they occupied. Yamīn al-Dawla Maḥmūd b. Sabuktegin was succeeded by his son Abū Saʿīd Masʿūd b. Maḥmūd. He dispatched an army against Mikāʾīl b. Saljūq's clan, who waged war, fought and defeated the Saljūqs. The victorious army seized several prisoners.

After Mikāʾīl's death, the Saljūqs and other Turkic groups gathered around his son Abū Ṭālib Muḥammad, nicknamed Ṭughril Beg. He was the eldest son of Mikāʾīl. Ṭughril, supported by his followers, fought against the Ghaznawid army of Masʿūd b. Maḥmūd. He defeated them soundly, and he killed and imprisoned many of its soldiers. Next he chased the withdrawing enemy up to Ṭūs (Mashhad) [in northern Iran], he attacked the city and conquered it. This was the first [Iranian] city captured by the Saljūqs. Ṭughril Beg and his supporters fortified the place and turned it to their bastion.

On the other hand, the sultan Masʿūd b. Maḥmūd fled to India. He travelled deep into the sub-continent and remained there for long years. The evacuation of Ghaznawid forces from Khurāsān inspired the Saljūqs and they endeavoured to conquer the region. Eventually they succeeded in taking over large parts of Khurāsān and in collecting taxes from its inhabitants. This roused the sultan Masʿūd to return to the battlefield. He fought against the Saljūqs, but was defeated by them. At that moment the ʿAbbāsid caliph al-Qāʾim bi-Amr Allāh dispatched messages to them, asking Ṭughril to protect the country and to guard the population. However, Masʿūd came back. Ṭughril inflicted heavy blows on the Ghaznawid army and scattered it. His grip on Khurāsān became firmly established (1035).

The year 431 (1039–40): In this year the sultan Masʿūd b. Maḥmūd prevailed over Ṭughril Beg b. Mikāʾīl b. Saljūq and routed him severely. He killed many of his men, captured a large number of prisoners and seized their weapons.

The year 432 (1040–41): In this year the sultan Ṭughril Beg b. Mikāʾīl b. Saljūq returned to Nishapur. The sultan Masʿūd b. Maḥmūd b. Sabuktegin had been defeated and he withdrew to Ghazna. Those Ghuzz tribesmen who were the followers of Ṭughril Beg seized the entire province of Khurāsān. The conquerors inflicted heavy blows on the urban population, killing and torturing. It is impossible to recount the suffering caused by them.

This year saw the execution of the sultan Masʿūd b. Maḥmūd b. Sabuktegin, the ruler of Khurāsān, Transoxiana and India. He was a brave king, a determined man, and a just leader, conversant with the rules of fighting and a good politician who took care of his subjects and lands. He was succeeded by his brother, the sultan Muḥammad b. Maḥmūd. His nephew Mawdūd was at that time in Ghazna. When he learned that his uncle had been put to death and that another uncle had taken hold of the reins of power, he assembled a force and travelled to India. He defeated his uncle Muḥammad and killed him. After that he returned to Ghazna and grasped the kingdom firmly.

In the year 435 (1043–44), an army of 1650 Ghuzz cavalry arrived. They were led by four commanders: Abū ʿAlī b. al-Dihqān, Gök-Bāsh (or Tāsh), al-Ḥājj Isrāʾīl and Abū Manṣūr. They ravaged and ruined the country of Āmid and Mayyafariqin, which at that date was controlled by the emir Abū Naṣr b. Marwān

(d. 1061). From these cities they headed to Nisibin. They cut the trees in its vicinity and besieged the city for a short period. Next they advanced southwards to Mosul. Its ruler, [Muʿtamad al-Dawla] Qirwash b. al-Muqallad [the ʿUqaylid], feared them, and hoping to buy peace, he agreed to pay them a considerable sum of money. Nevertheless, they devastated the province of Diyār Bakr (Diyarbakır) and actually the entire Jazira (northern Iraq–southern Turkey). Following this, the Ghuzz moved to Mosul and attacked the city, killing, imprisoning, and wreaking havoc until they actually conquered the region. The Ghuzz ruled over these places for a short period.[350]

Yet the Arabs managed to regroup and attacked the Ghuzz. They besieged them in Mosul and fought against them till they forced the Ghuzz out of there. The fighting did not end, and the parties continued to exchange blows. The Ghuzz were killed and the Arabs won. The Ghuzz fled to Mayyafariqin. They plundered where they could and continued this till they found refuge in Azerbayjan.

In this year (435/1044) the ruler of Baghdad, the king Jalāl al-Dawla Abū Ṭāhir Fayrūz, the son of the king Bahāʾ al-Dawla b. ʿAḍud al-Dawla b. Būyah (Buwayh), passed away.[351]

In the year 442/1050–51 the ruler of India, the sultan Mawdūd b. Masʿūd b. Muḥammad b. Sabuktegin passed away. His uncle ʿAbd al-Rashīd b. Maḥmūd inherited the kingdom.

In this year the ruler of Mosul the emir Muʿtamad al-Dawla Qirwash b. al-Muqallad died. He was a brave sovereign and a gifted leader.

In the year 447/1055 Ṭughril Beg arrived in Baghdad and the reign of the Saljūq dynasty commenced. The circumstances of these events were as follows: al-Muẓaffar ("the victorious") Abū al-Ḥārith Arsalān the Turk, who is known by the nickname al-Baṣāṣīrī, had become a strong and forceful power in Iraq. He seized the country and governed it. His fame spread and his image built up. Rival Persian and Arab commanders feared him. His name was mentioned at the Friday sermons (*khuṭba*) in the mosques and he was recognized as the *de facto* ruler of Iraq. Al-Malik al-Raḥīm (the Merciful King), the *de jure* ruler, was left with only the title.

Following these developments, al-Malik al-Raḥīm informed the caliph al-Qāʾim bi-Amr Allāh that al-Baṣāṣīrī intended to raid the caliph's palace, to plunder it and to arrest the caliph. Consequently al-Qāʾim bi-Amr Allāh wrote to the sultan Ṭughril Beg b. Mikhāʾil b. Saljūq, who was at that time in the province of al-Rayy, calling him to come and help him, and inviting him to travel to Iraq. We have already mentioned above the emergence of Ṭughril Beg and the accounts of the conquest of Khurāsān by him.[352]

At that time, al-Baṣāṣīrī, accompanied by his supporters, was camped in al-Wāsiṭ in southern Iraq. Some of his men deserted him, as a result of these developments, and returned to Baghdad. They plundered al-Baṣāṣīrī's palace and set it on fire. Meanwhile the sultan Ṭughril Beg arrived at Baghdad (in December 1055). Al-Baṣāṣīrī camped on the banks of the Euphrates and controlled the territory up to the city of al-Raḥba in the north. He wrote to Cairo, to the Fatimid caliph al-Imām al-Mustanṣir bi-Allāh, the ruler of Egypt, acknowledging his authority. Henceforth al-Mustanṣir's name was mentioned in the mosques of al-Raḥba[353] during the ceremonial prayers. In response al-Imām al-Mustanṣir sent money to al-Baṣāṣīrī.[354]

When the sultan Ṭughril Beg arrived in Baghdad his name was mentioned in the Friday sermon immediately after the name of the ʿAbbāsid caliph. The name of the Buyid (Buwayhid) prince al-Malik al-Raḥīm followed Ṭughril Beg's name. This resulted from the caliph's appeal to the sultan and his intercession on behalf of the Buyid prince. However, after a short period Ṭughril Beg ordered the arrest of al-Malik al-Raḥīm. In the last days of Ramaḍān 447 (December 1055) the name of the Buyid prince was removed from the *khuṭba.* This signified the end of the Buyid dynasty and the extinction of their rule. They governed for one hundred and twenty seven years. Allah, Whose rule never ends, be Praised. Praise the Nominator of kings, the One who gives sovereignty unto whom He wishes, and the One who withdraws sovereignty from whom He wishes. He is the originator of the nations and their restrainer. He is the depriver of favours and their restorer.[355]

The sultan Ṭughril Beg brought eighteen elephants with his army to Baghdad. He occupied the government's place and grasped firmly the reins of power.

The caliph al-Qāʾim bi-Amr Allāh married the lady Arsalān Khatūn Khadīja, the daughter of Dāʾūd b. Mikāʾīl b. Saljūq, the sister of the sultan Ṭughril Beg (read: his brother's daughter), and paid the sum of one hundred golden dinars as bride money (this happened in the year 458/1056).[356]

In this year, after camping in Baghdad for the period of thirteen months and thirteen days, the sultan Ṭughril Beg travelled north from Baghdad toward the province of Mosul. His soldiers carried with them mangonels and ballistae. He first approached Tikrit and besieged the city. Disturbances broke out during this period in Kūfa, Wāsiṭ and the oasis of ʿAyn Tamar (lit. the Date Spring). In these places the name of the Fatimid caliph al-Imām al-Mustanṣir bi-Allāh was mentioned in the *khuṭba.*

In the year 449/1058 the caliph al-Qāʾim bi-Amr Allāh bestowed seven robes of honour on the sultan Ṭughril Beg.[357] The caliph also adorned him with a ring on his neck, put a crown on his head and ornamented him with bracelets on his arms. In addition he handed him a written nomination certificate. Ṭughril's authority over all the regions that stretch from the gate of the caliph's palace and beyond was recognized and his authority became firmly established. No one contested him, neither in Khurāsān nor either side of the Tigris River.

The name of the caliph al-Qāʾim bi-Amr Allāh had been removed from the *khuṭba* and replaced by the name of al-Mustanṣir bi-Allāh, the Ismāʿīlī ruler (*imām*) of Egypt. This happened during Ṭughril Beg's absence from Baghdad. The sultan departed the city and advanced northwards towards the city of Mosul. From there he proceeded to Nisibis. He sought to conquer these territories and to annex them to his domain. He was accompanied by his brother Ibrāhīm Yināl. The reason that al-Mustanṣir bi-Allāh's name was mentioned during the sermons in Baghdad's mosques was because al-Baṣaṣīrī revolted against the ʿAbbāsid caliph.

Al-Baṣaṣīrī had dispatched messengers to Ibrāhīm Yināl and encouraged him to rebel against his brother Ṭughril Beg. He promised to help him and tempted him to seize the kingship. Ibrāhīm Yināl recruited soldiers and, leading a sizeable army, departed Iraq and proceeded to al-Rayy in Iran. He rebelled openly against his brother Ṭughril Beg. The sultan advanced towards him and the two armies clashed. Ibrāhīm won the battle and blockaded Ṭughril Beg, who had fortified himself in the city of Hamadan.[358]

The Lady (*khātūn*), Ṭughril Beg's wife, decided to rescue him and her steps caused unrest in Baghdad. Rumors about the advance of al-Baṣāṣīrī towards the caliph's capital intensified the panic. Al-Kundurī and Anushirvān stopped proceeding towards Hamadan. The *khātūn*, who disagreed with them about the measures to be taken, contemplated arresting them. Al-Kundurī and Anushirvān took refuge on the western bank of the Tigris River and destroyed the bridge that connected the two banks of the river. The Ghuzz mutinied and plundered the dwellings of the *khātūn*. They grabbed everything. Baghdad became empty of armies and anarchy prevailed in the city.

Al-Baṣāṣīrī entered the city on Sunday, 8 Dhū al-Qaʿda 450/27 December 1058.[359] His army carried the banners of the Egyptian *imām* al-Mustanṣir bi-Allāh. His name and title "the Commander of the Faithful Abū Tamīm Maʿadd" were written on the flags. The inhabitants of al-Karkh quarter[360] inclined towards him. They changed the call to prayer to the Shīʿī version: "Come and do the best of work." Other flocked to the caliph al-Qāʾim bi-Amr Allāh. Sailing on boats in the Tigris, the parties fought each other.

On Friday (13 Dhū al-Qaʿda 450/1 January 1059) the *khuṭba* in the cathedral mosque that al-Manṣūr had built in Baghdad was made in the name of the Egyptian *imām* al-Mustanṣir bi-Allāh. In the call to prayer the line "Come and do the best of work" was added. The bridge was rebuilt. Al-Baṣāṣīrī's forces crossed it to the eastern bank of the Tigris. The name of al-Mustanṣir bi-Allāh was mentioned also in the sermons in the suburb of Ruṣāfa.[361] The caliph al-Qāʾim bi-Amr Allāh had ordered to dig trenches around his palace and sheltered behind them while the enemy closed on him. During the fighting the mob burned the region of al-Muʿalla canal (*nahr*)[362] and pillaged it, snatching large quantities of goods. Al-Baṣāṣīrī won the day and most of the caliph's supporters deserted him.

The caliph al-Qāʾim bi-Amr Allāh sought refuge with Quraysh b. Badrān. This ʿUqaylid commander joined forces with al-Baṣāṣīrī, yet nevertheless he provided shelter to the caliph and his followers. Al-Baṣāṣīrī agreed with this.[363] But when he captured the vizier Raʾīs al-Ruʾasāʾ b. al-Muslima, the secretary of the caliph, he bound him in chains and dressed him with a woolen outer garment and a tall bonnet of red felt, and around his neck was a collar of camel skin. Seated on a camel, the vizier was paraded in the streets of Baghdad. Behind him was seated a man who slapped him constantly all the way, till they arrived at a wooden gibbet that had been erected for him. He was taken and dressed in the skin of an ox that had been slaughtered that very day. The ox's horns were put on the vizier's head. He was raised with iron hooks and beaten violently till he died.[364]

The ʿAbbāsid caliph al-Qāʾim bi-Amr Allāh continued his journey and from the camp of Quraysh b. Badrān he moved to Muhārish's camp on the eastern bank of the Tigris River. There, a tent had been erected to house him. Meanwhile a huge crowd sacked the caliphal palace. The mob carried away large quantities of objects from it.

The mosque of the caliph had been deserted. No prayer took place in it and on Friday (4 Dhū al-Ḥijja 450/22 January 1059) no one preached in it and no *khuṭba* was said. In the other mosques of Baghdad, the name of the Ismāʿīlī *imām* al-Mustanṣir bi-Allāh was pronounced. The mention of the ʿAbbāsid caliph's name during the religious sermons came to an end.

On that very day al-Qāʾim bi-Amr Allāh reached Ḥadīthat Ana, where he was entrusted to the local ruler Muhārish b. Mujallá, who put the caliph under house arrest. That happened because al-Baṣāṣīrī and Quraysh b. Badrān disagreed about the caliph's fate. So they decided that al-Qāʾim bi-Amr Allāh would be lodged with Muhārish until they reached a solution.

Al-Baṣāṣīrī ordered (in 451/1059) the chief judge (*qāḍī*) Abū ʿAbd Allāh [Muḥammad] al-Dāmghānī [al-Ḥanafī al-Kabīr] (1007–85), other judges, the preachers, and the descendents of Hāshim, the great-grandfather of the Prophet, be brought to his encampment. There they were forced to pay obeisance to *imām* al-Mustanṣir bi-Allāh, the ruler of Egypt. He extracted an oath of allegiance from them.

In this year (451/1059) the sultan Ṭughril Beg defeated his brother Ibrāhīm Yināl and overpowered him. He [ordered] Ibrāhīm be strangled with a bowstring. Next he received the allegiance of those Turkoman tribes who had supported Ibrāhīm Yināl's uprising against him. After the Sultan had settled this affair, he recovered his position. He strengthened his standing. Then he started back to Baghdad, intending the killing of al-Baṣāṣīrī and the restoration of the caliph al-Qāʾim bi-Amr Allāh to his palace.[365]

The sultan Ṭughril Beg sent the sheikh Abū Bakr b. Furak[366] to the caliph al-Qāʾim bi-Amr Allāh to provide for the caliph's needs. He carried gifts and coats [forty different types of clothes and sixty tailored robes], worth five thousand golden dinars. In addition, [Ṭughril Beg's wife] sent with the messenger sixty-five tailored women's robes to the wife of the caliph al-Qāʾim bi-Amr Allāh.

While the Sultan was advancing toward Baghdad, Muhārish the ruler of al-Ḥadītha, along with the caliph al-Qāʾim bi-Amr Allāh, set out on 11 Dhū al-Qaʿda [451/9 December 1059] to meet Ṭughril Beg. Near Qalʿat Tall ʿUkbara[367] they met the master (*ustadh*)[368] Abū Bakr b. Furak, the ambassador of the sultan Ṭughril Beg. Ibn Furak delivered [the sultan's message] and gifts and they pleased the caliph very much.[369]

During Ṭughril Beg's arrival in Baghdad his army plundered the city. Not only was the suburb of al-Karkh pillaged by these soldiers, but the entire city was looted. Ṭughril Beg ordered great quantities of tools, boats and tents be produced, and sent them to the Caliph. [Ṭughril selected from among the royal guards a contingent of three hundred servants who led fourteen Bactrian camels carrying a gigantic pavilion, and in addition tents and yurts, as well as tools and carpets, a mule carrying a veiled cradle and three horses with gold saddles.] The heads of the mission were the vizier ʿAmīd al-Mulk al-Kundurī and the *ustādh* Abū Bakr b. Furak. A large tent had been erected to house the Caliph, who dwelled in it for a couple of days.[370] Afterwards he continued towards Baghdad. A camp of pavilions was set up to accommodate him.

Leading his armies, Ṭughril Beg rode to Baghdad to meet the caliph. The moment he saw the caliph's pavilion, he dismounted from his horse and kissed the ground seven times. The Caliph had taken a pillow from his throne and threw it towards the Sultan, who grasped it and kissed it before sitting on it. Ṭughril Beg pulled out from his gown the red corundum[371] that previously had belonged to members of the Buyid dynasty and twelve unique pearls and presented them to the caliph al-Qā'im bi-Amr Allāh. He stated that this was a gift from the Lady Arsalān Khatūn, the wife of the Caliph. The Sultan apologized for his own tardiness on account of Ibrāhīm Yināl's rebellion, due to which he had been slow to return to Baghdad.

The Caliph continued his journey and entered Baghdad. The sultan Ṭughril Beg walked ahead him, holding the bridle of the caliph's mule and led it until he reached the al-Ḥajra (inner apartments) Gate. His arrival was on [Monday] 25 Dhū al-Qaʿda [451/3 January 1060). It was the very same day a year earlier that the Caliph departed his palace. He was away from his capital for exactly a year.

Al-Baṣāṣīrī considered the situation a petty development and a couple of days prior to the Caliph's return to Baghdad had moved southward to Wāsiṭ. There he was busy with the collection of farming products and with loading them on riverboats. He planned to carry the foodstuffs northwards up the Tigris to Baghdad. Yet, when he learned that the Caliph and the Sultan had entered Baghdad he advanced northward to al-Nuʿmaniyya,[372] accompanied by a small flotilla of boats. Ṭughril Beg sent an army (at the end of January 1060), led by Khumar-Tegin [al-Tughrā'ī][373] and a team of commanders [towards the opponent's camp]. The Sultan planned to follow them soon with his forces. Fighting erupted and al-Baṣāṣīrī was killed. His head was carried to the Sultan. It was put on a lance and paraded along the streets of Baghdad in Dhū al-Qaʿda [read Dhū al-Ḥijja] 451/ January 1060.[374] The Caliph returned to his palace and order was restored in the streets of Baghdad.

When he was seized, the caliph al-Qā'im bi-Amr Allāh had sworn that [if released] he would personally prepare his breakfasts, pardon those who misbehaved and forgive those who attacked him. The Sultan went down to Wāsiṭ and reorganized the city's administration, due to which the population was calmed.[375] [The sultan ordered the] transport to Baghdad of the Caliph's mother and her maids [who had been captives of al-Baṣāṣīrī]. [Ṭughril Beg] sent orders by the postal service concerning al-Baṣāṣīrī's [womenfolk and children].

The sultan Ṭughril Beg returned to Baghdad in the year 452/1060. The Caliph gave him robes of honour and demonstrated great respect towards him. Afterwards [in April 1060] he went to the uplands in western Iran (Jibāl).[376] He had ordered his vizier ʿAmīd al-Mulk al-Kundarī to serve on his behalf in Baghdad. The Sultan returned to the city after he had finished reforming the administration of the Uplands (Jibāl). His circumstances improved considerably.

In the year 453/1061 the sultan Ṭughril Beg was contracted to marry the daughter of the caliph al-Qā'im bi-Amr Allāh. The Caliph was disturbed by this and declined. Messengers went to and fro and threats were exchanged. Long explanations would be needed to elucidate the haggling. The Caliph persisted in refusing and made plans to leave Baghdad. He ordered to prepare a fast sailing ship (*ṭayyāra*)[377] and to equip it, to polish, clean and repair it. Threats and promises by the men of the Sultan did not stop, nor threats that would infringe the Caliph's immunity.[378]

Eventually the sultan Ṭughril Beg ordered, in the year 453/1062, the removal of the Caliph's agents from the privy budget. When the Caliph was told the news, he was advised to respond favourably to the Sultan's proposal to marry his daughter. The conflict dragged on for many months. The Caliph could not make up his mind and changed positions, till eventually he answered positively. Although resenting the proposal, he gave his agreement to drawing up a marriage document and to sending it to Ṭughril Beg [who was camped outside Tabriz]. The Caliph assigned [ʿAmīd al-Mulk al-Kundurī] the power to act as his proxy. The Sultan Ṭughril Beg was very pleased with this and ordered to restore all the farming lands that were the property of the Caliph and his entourage, which he had seized. He dispatched large sums of money and precious gifts to the caliph al-Qāʾim bi-Amr Allah in 454/1062.[379]

<A marginal note> It was narrated that the sultan Ṭughril Beg saw himself in a dream as though he was carried up to heaven. He was told: "Ask whatever you wish." The Sultan responded: "I request long life." The reply was: "You will live seventy years", and indeed he did not live longer.

In [Muḥarram] 455/January 1063, the Sultan Ṭughril Beg travelled to Baghdad [from Urmiyya]. The wedding procession then transferred his bride, the daughter of the caliph al-Qāʾim bi-Amr Allāh, to him. He showed signs of great joy. He placed [considerable gifts] before her. He bestowed robes upon all the eunuchs, servants, maids and pages, who accompanied her. He also bestowed robes on the Caliph's young slave troops and showed great respect towards them.

A short time later [March 1063] the Sultan Ṭughril Beg became seriously ill. He was carried from Baghdad to Hamadan. His wife, the daughter of the caliph al-Qāʾim bi-Amr Allāh, accompanied him. Although first she rejected the idea of travelling with him, he however insisted that she go with him.

During this year, on Friday, 8 Ramaḍān 455/5 September 1063, the Sultan died in Rayy. He ruled for thirty years and was aged seventy. His marriage with the Caliph's daughter lasted six months and several days. Ṭughril Beg was a wise and tactful king, one of the best rulers. He generated fear and respect in the hearts [of all]. Kings admired him, corresponded with him and dispatched gifts to him. When his predetermined days on Earth arrived at their limits, pre-destined death approached and carried him from his roomy palace to his narrow grave. Blessed is God the eternal and everlasting whose kingdom never will come to an end.

After the death of the Sultan Ṭughril Beg, his nephew the sultan ʿAḍud al-Dawla Alp-Arsalān Muḥammad, the son of Dāʾūd [Chaghri Beg] b. Mikāʾīl b. Saljūq, sat upon the throne. The *khuṭba* in Baghdad mentioned him and the Caliph dispatched to him a letter containing his investiture as sultan and banners. Alp-Arslān nominated Niẓām al-Mulk Abū ʿAlī Ḥasan b. Isḥāq al-Ṭūsī to serve as his vizier.[380]

b) Ibn al-Dawādārī: The history of the Turks

In this year (425/1033–34) Saljūq advanced towards Transoxiana and Bukhara. He clashed with the armies of Maḥmūd. This is the famous Dāghān (sic; might refer to Dāndanqān in 431/1040) campaign. The armies of Maḥmūd b. Sabuktegin crumbled. He decided to lead his armies personally but death reached him before he did so. We mentioned it already previously. Hence the rule of Saljūq grew stronger.[381]

In this year (431/1039–1040) the emissaries of the great *khān* arrived at Ṭughrīl Beg's camp, bringing with them fine gifts. With them came a man with no head and no neck. His face was at the chest, his eyes resembled wicker, his mouth was at his bosom. He was accompanied by a translator who comprehended his tongue. He reported to Ṭughrīl Beg who asked him about his origins. Via the translator, he replied that he was from the land of the Qarāṭāgh (Turkish *kara dağ* = the black mountain)[382] and that he belonged to a great people, whose multitudinous numbers only the Almighty knows. They live on the shores of the Ocean. They have no king and no religion. They resemble wild animals and possess no knowledge which humans acquire. They are the progeny of Japheth the son of Noah. Ṭughrīl Beg turned to the translator and asked him about this person. He said: We call this people *bashi baq ughli* (Turkish: *baş baq uğlu*), which means "the headless son."

It is narrated, in regarding to their origins, that in remote times their father was beheaded unjustly, yet his corpse continued to live and he did not die. He walked and moved. If he became hungry a vein popped up between his shoulder and it was recognized that he was hungry. Food and drink were provided to him. All his sons were born in this shape. A distance of a difficult two years' [journey] across lands, mountains, rivers and sands separates their lands and the territory of the khān.

The king Gumush (Turkish: *gümüş,* silver) Khān, the son of the great prince (*al-ṭarkhān al-kabīr*), whose kingdom lies opposite the kingdom of the khān, has sent this person, among other gifts to the khān, to amuse him. The khān sent him to amaze you [Ṭughrīl] with the Creator's powerful might.[383]

The Saljūqids: Their origin and early history

The author of *The History of Baghdad* says:[384] The Saljūq household was Turkmen, living in camps and tents in the steppes of Transoxiana. Saljūq married a daughter of Yakar-Khān or Duquz-Tekin,[385] which probably is the correct form, who was one of the Turkmens' kings. In those days Maḥmūd b. Sabuktekin[386] was the supreme ruler of this vast abode. Saljūq's strength increased as result of this royal marriage and he was able to destabilize Maḥmūd's authority in Transoxiana. He became the head of many brave and combative Turkmens.

Originally Saljūq was a scion of a leading family. It is said that his royal family was descended from the imperial Sasanid house, the kings of Persia. This data was narrated by the author of *The History of Baghdad*, yet he did not produce any evidence to back up his tale and restricted his account to these limited words.

Evidence that was extracted from other sources and which demonstrates the accuracy of the story that Saljūq indeed was a scion of the Sasanid house will be presented in another place.

The author of *The History of Baghdad* continues the account and says that being aware of his success in building up a powerful force, Saljūq decided to conquer the neighboring territories. His royal aspirations motivated him to destabilize the Ghaznawid realm and to challenge their authority. Maḥmūd b. Sabuktekin decided to take to the field and to eliminate the challenge, yet he passed away. Death reached him before he was able to attack Saljūq. The kingdom fell into the hands of Masʿūd b. Maḥmūd, who was at a tender age, inexpedient and childish. This development was according to the Lord's will, who bestowed majesty on the house of Saljūq. They used to camp in four thousand tents.

Ṭughril Beg, the son of Mikāʾīl b. Saljūq, matured and many Turkmens and other people adhered to him. He advanced against Masʿūd and encamped in Nishapur. As Ṭughril Beg went from strength to strength and his authority built up, the companies of Masʿūd broke away. Masʿūd fled from him and deserted the land. Ṭughrīl Beg seized Khurāsān peacefully with no fighting or war and became the country's ruler. In the year 431/1040 he ascended the throne. I hope to elaborate on the Turks' history further below, if Allah would permit it. This piece of information on the emergence of this people is sufficient at this point. I will produce a detailed account of their story based upon different narratives.[387]

Notes

1 Ibn Khurdādhbih, *Kitāb al-Masālik wal-mamālik* (*The Book of Itineraries and Kingdoms*), ed. M. De Goeje (Leiden, 1889; reprinted 1967), 30–1, 261–3; trans. *idem* (Leiden, 1889), 202–4.

2 Adam J. Silverstein, *Postal Systems in the Pre-modern Islamic World* (Cambridge, 2007), 64.

3 ʿAbd al-Malik b. Muḥammad b. Ismāʿīl al-Thaʿālibī (961–1038), *Histoire des rois des Perses*, trans. H. Zotenberg (Paris, 1900), 130.

4 Originally Qaghan or Khaghan. A Turkic title meaning supreme ruler; A.G.C. Savvides, "Some Notes on the terms Khan and Khagan in Byzantine Sources", in I.R. Netton (ed.), *Studies in Honour of Clifford Edmond Bosworth* (Leiden, 2000), 1: 267–79.

5 The nine (*dokuz* in modern Turkish) Oghuz tribes; T. Senga, "The Toquz Oghuz Problem and the Origin of the Khazars", *Journal of Asian History* 24/1 (1990): 57–61.

6 A unit of length used in ancient Persia, originally the distance travelled in an hour; also transcribed as *farsang* or *farsakh*. Silverstein, *Postal Systems*, 93–4.

7 Compare with the account of the Toghuz-Ghuz in Qudāma ibn Jaʿfar Abū al-Faraj al-Khaṭīb al-Baghdādī (d. 948), *al-Kharāj wa-ṣināʿat al-kitāba*, ed. M.H. al-Zubaydi (Baghdad, 1981), 196 (ll. 1–4, 9): "In the city of the Toghuz-Ghuz there is a lake, which is surrounded by villages and clusters of hamlets. [There are] twelve iron gates to the city. All the Turkic people guard the place."

8 Talas in contemporary Kazakhstan.

9 This ethnonym may be related to the Turkish word *kar* (snow). In the eighth century, the Qarluq advanced to the western Turkic lands and established political relations with the Muslims; P. Golden, *An Introduction to the History of the Turkic People:*

Ethnogenesis and State-Formation in Medieval and Early Modern Eurasia and the Middle East (Wiesbaden, 1992), 196–9.

10 See Ibn Faqīh below.

11 Michael Bonner, "The Naming of the Frontier", *BSOAS* 57/1 (1994): 17–24.

12 Le Strange, *The Lands of the Eastern Caliphate*, 376–81 (and map 5).

13 Francis Clark Murgotten (trans.), *The Origins of the Islamic State: Being a Translation from the Arabic, Accompanied with Annotations, Geographic and Historic Notes of the Kitab futuh al-buldan* (New York, 1924), part 2: 41–2; C.E. Bosworth (trans.), *The History of al-Tabari* vol. 5: *The Sasanids, the Byzantines, the Lakhmids, and Yemen* (New York, 1999), 112–13 (and notes); and see below.

14 Minorsky (trans.), *Ḥudūd al-ʿĀlam*, 97–8, 286–97.

15 Vladimir Minorsky, "Tamim ibn Bahr's Journey to the Uyghurs [in *c.* 821 AD]", *BSOAS* 12 (1948): 280, 283 (item 7 and 8), 290–1; and my translation of Ibn Faqīh.

16 Minorsky (trans.), *Ḥudūd al-ʿĀlam*, 99–100, 304–10.

17 C.E. Bosworth, "al-Xwarazmi on the people of Central Asia", *JRAS* (1965): 4–5.

18 Minorsky, "Tamim ibn Bahr's Journey", *BSOAS* 12 (1948): 284 (item 12).

19 Le Strange, *The Lands of the Eastern Caliphate*, 486–7.

20 This translation is based on two editions of the Mashhad manuscript 5529 (discovered in 1923 by Ahmed Zeki Validi Togan): 1) Abū Bakr Aḥmad b. Muḥammad b. Isḥaq Ibn al-Faqīh al-Hamadhāni (fl. 902), *Nuṣūṣ lam tuḥaqqaq min kitāb akhbār al-buldān* eds. al-Zahrānī and al-ʿAsīrī (Mecca, 1417/1997); 2) *Kitāb al-Buldān* ed. Y. al-Hādī (Beirut, 1996). This new edition supercedes the abridgement published by M. de Goeje (Leiden, 1885) and translated by Henri Massé, *Abrégé du livre des pays* (Damascus, 1973).

21 Aloys Sprenger, *Die Post- und Reiserouten des Orients* (Leipzig, 1864), introduction, XVII–XVIII.

22 Paul Kahle, "Islamische Quellen zum chinesischen Porzellan", *ZDMG* 88 (1934): 12, 43–5 (on Ibn al-Faqīh); Otar Tskitishvili, "Concerning the Tashkent MS of a Geographical Treatise by Ibn al-Faqīh al-Hamadānī", *JESHO* 23/3 (1980): 304–8; Zacharias van Laer, "Ibn al-Faqīh et le Livre des pays", *Arabica* 31 (1984): 69–79.

23 Y. Frenkel, "The Turks of the Eurasian Steppes in Arabic Geographical Literature", in M. Biran and R. Amitai (eds.), *Mongols, Turks and Others: Eurasian Nomads and the Sedentary World* (Leiden, 2005), 225.

24 Cf. Bashear, *Arabs and Others in Early Islam,* 104–8.

25 This paragraph opens with the verb "he said". It closely follows Ibn Khurradādhbih, see section A above; additional information on the Turks of Inner Asia in the tenth century is provided by Minorsky (trans.), *Ḥudūd al-ʿĀlam*, 263–317.

26 The Oghuz were a political tribal organization who appeared at various times and places in the Eurasian steppes. There are several theories regarding the name's etymology. One suggests it is derived from *oq* "arrow" that has taken on the meaning "tribe". The ethnonym Toghuz-Oghuz can be translated as the nine tribes of the Oghuz. Peter B. Golden, "The Migrations of the Oghuz", *Archivim Ottomanicum* 4 (1972): 45–84 (reprinted in his *Nomads and their Neighbours in the Russian Steppe* (London, 2003)).

27 In the ninth century they roamed Kazakhstan. Peter B. Golden, "The Peoples of the South Russian Steppes", in Denis Sinor (ed.), *The Cambridge History of Early Inner Asia* (Cambridge, 1990), 278–80; *idem*, *An Introduction to the History of the Turkic People*, 202–5.

28 A sub-confederation of the Qarlug centred in Mongolia.

29 In the ninth century, the Pechenegs were active in the middle Syr Darya region. In the tenth century, they moved westward to the lands of present-day southern Russia. P. Golden, "The Peoples of the South Russian Steppes", 270–5.

30 This Turkic people moved during the tenth century from southern Siberia towards the Volga River. P. Golden, "The Qipcaqs of Medieval Eurasia: An Example of Stateless

Adaptation in the Steppes", in G. Seaman and D. Marks (eds.), *Rulers from the Steppe: State Formation on the Eurasian Periphery* (Los Angeles, 1991), 133–4 (reprinted in his *Nomads and their Neighbours in the Russian Steppe* (London, 2003)).

31 Minorsky (trans.), *Ḥudūd al-ʿĀlam*, 94–101, 264–5 (commentary).

32 On this term, see Irfan Shahid, *Byzantium and the Arabs in the Sixth Century* (Washington, DC, 2002), 2/1: 66, 176, 223.

33 Minorsky (trans.), *Ḥudūd al-ʿĀlam*, 118, 358 (note 88 Farab or Parab).

34 That is, "they hold an extremely important position". But see in al-Masʿūdi, *Kitāb al-tanbīh wal-ishrāf*, ed. M.J. De Goeje (Leiden, 1893), 180 (l. 8) "nomadic Turkic groups (*ajnās min al-turk badw*[un])".

35 The square brackets indicate clauses not found in Ibn Faqīh's extant text, but quoted in other sources, such as Yāqūt, *Muʿjam al-buldān*, *s.v.* "Turkistan".

36 Ibn Faqīh describes these distances and roads also in the chapter on Khorasan. Le Strange, *Eastern Caliphate*, 488–89.

37 Called Chach by the Persians. Its ruins are now known as Old Tashkent. Minorsky (trans.), *Ḥudūd al-ʿĀlam*, 118 (no. 79), 357 (commentary); Le Strange, *Eastern Caliphate*, 480–2.

38 M.S. Asimov and C.E. Bosworth (eds.), *History of Civilizations of Central Asia: Age of Achievement: A.D. 750 to the End of the Fifteenth Century* (Paris, 1998), 4/1: 74–7; Mehmet Fuat Köprülü, *Early Mystics in Turkish Literature*, trans. and ed. Gary Leiser and Robert Dankoff (London, 2006), 148 (16), 155 (29).

39 Max Meyerhof, "ʿAlī at-Ṭabarī's 'Paradise of Wisdom', one of the oldest Arabic Compendiums of Medicine", *Isis* 16 (1931): 7–11; Konstantin A. Inostranzev, *Iranian Influence on Moslem Literature* (part 1 translated with supplementary appendices from Arabic sources by G.K. Nariman) (Bombay, 1918), 93–4.

40 A repeated sentence, see above, p. 636.

41 C.E. Bosworth (trans.), *The Laṭāʾif al-maʿārif of Thaʿālibī* (Edinburgh, 1968), 142.

42 This nickname indubitably refers to Tamīm's career as a Muslim volunteer in the Eurasian frontier (*ribāṭ*). His Eurasian journey took place in the first quarter of the ninth century. Several Arab authors quote from his travelogue, which was published and studied by Vladimir Minorsky as "Tamim ibn Bahr's Journey to the Uyghurs", *BSOAS* 12 (1948): 275–305 (reprinted *idem*, *The Turks, Iran, and the Caucasus in the Middle Ages* (London, 1978), art. 1) (see further below).

43 The following paragraph actually repeats Tamīm's account. Minorsky, "Tamim ibn Bahr's Journey", *BSOAS* 12 (1948): 283 (1).

44 Ibn Faqīh, ed. de Goeje, 329 (l. 10): "The dominant doctrine among the Turks is the dualistic heresy (*zanādiqa*)."

45 So in Yāqūt, but only one hundred persons in Ibn Khurdādhbih and Ibn Faqīh (ed. al-Hādī, 627, ll. 20–1).

46 This section is a repetition of a previous description. Additional sources are referred to in the notes.

47 This paragraph is repeated below: "He says that its length is a journey of forty days on camels, but a horseman can cover the distance in a month if he rides hard." Minorsky, "Tamim ibn Bahr's Journey", 284 (item 10). In al-Hadi's edition this repetition is omitted.

48 Ibid., 284 (item 9).

49 Ibid., items 11–12.

50 Qudāmā ibn Jaʿfar, *al-Kharāj*, 105 (note 239; the editor changed the name to Kawakit).

51 Minorsky, "Tamim ibn Bahr's Journey", 285 (13).

52 An astronomer from Ghazna. A. Akhmedov, "Astronomy, Astrology, Observatories and Calendars", in Asimov and Bosworth (eds.), *History of Civilizations of Central Asia* (Paris, 2000), 4/2: 196–7.

53 W. Barthold (Bartold), *Turkestan down to the Mongol invasion,* 3rd edn (London, 1968/1977), 166, 201–2.

54 Tsien Tsuen-Hsuin, *Chemistry and Chemical Technology* pt. 1 *Paper and Printing* (Cambridge, 1985), 297.

55 Minorsky reads b. Lohrāsp in al-Hadi's edition; see next note.

56 According several Arab-Muslim authors, some Turkic peoples were related to the Arabs by being the offspring of Abraham, the father of Ishmael. According to this imaginary ethnic history, three sons of Qanṭūra, Abraham's wife, mingled with the Khazar Turks. The Banū Qanṭūra were said to have wide faces and narrow eyes. William M. Brinner (trans.), *The History of al-Tabari* vol. 2: *Prophets and Patriarchs* (Albany, 1987), 129.

57 Biblical Keturah in Genesis, 25: 1–5; G.D. Newby, *The Making of the Last Prophet: A Reconstruction of the Earliest Biography of Muhammad* (Columbia, SC, 1989), 79.

58 The Banū Qanṭūra played a role in some apocalyptic traditions. Geert Jan van Gelder, "Mirror for Princes or Vizor for Viziers: The Twelfth-Century Arabic Popular Encyclopedia '*Mufīd al-ʿulūm*' and Its Relationship with the Anonymous Persian '*Baḥr al-fawa'id*'", *BSOAS* 64 (2001): 334.

59 Midyan is often linked by Muslim commentators to the Midianite people, the prophet Jethro and the people of al-Ayka (the men of the thicket) who are mentioned in several Qur'anic verses. C.E. Bosworth, "Madyan Shu'ayb in Pre-Islamic and Early Islamic Lore and History", *Journal of Semitic Studies* 29 (1984): 53–64; Isaac Hasson, "Judham entre la Jahiliyya et l'Islam", *Studia Islamica* 81 (1995): 13–15.

60 The Arab tribe of Judham is associated in some traditions with Medan (Madan). I. Hasson, "Judham entre la Jahiliyya et l'Islam", *SI* 81 (1995): 9–10.

61 Israel Eph'al, *The Ancient Arabs: Nomads on the Borders of the Fertile Crescent 9th–5th Centuries B.C.* (Jerusalem, 1982), 232–3.

62 An Armenian tradition says: "Kendura (i.e. Ketura), the handmaiden of Abraham, bore Imran who became *the ancestor of* the Parthians, from whom came the brave Arshak and St. Gregory the Illuminator of Armenia." Richard Nelson Frye and Robert Blake, "History of the Nation of the Archers (The Mongols) by Grigor of Akanc; Hitherto Ascribed to Magakia the Monk: The Armenian Text edited with an English Translation and Notes", *Harvard Journal of Asiatic Studies* 12 (1949): 287.

63 See Minorsky, "Tamim ibn Bahr's Journey", 285 (15).

64 al-Tahamani. An official (d. 910) in the service of Ismāʿīl b. Aḥmad, the Samanid ruler of Transoxania (fl. 892–907). John Andrew Boyle, "Turkish and Mongol Shamanism in the Middle Ages", *Folklore* 83 (1972): 187–8; *idem*, "The Alexander Legend in Central Asia", *Folklore* 85 (1974): 220–1.

65 Members of the Bādhghisī family served in the ʿAbbāsid administration of Armenia. C.J.F. Dowsett, "A Neglected Passage in the *History of the Caucasian Albanians*", *BSOAS* 19/3 (1957): 457; C.E. Bosworth (trans.), *The History of al-Tabari:* vol. 32 *The Reunification of the ʿAbbāsid Caliphate* (Albany, 1987), 25.

66 See the employment of this term in al-Jāḥiẓ. W.M. Hutchins, *Nine Essays of al-Jāḥiẓ* (New York, 1975), 179–80.

67 Richard Nelson Frye and Aydin M. Sayili, "Turks in the Middle East before the Saljūqs", *JAOS* 63 (1943): 199; D.G. Tor, "The Islamization of Central Asia in the Sāmānid Era and the Reshaping of the Muslim World", *BSOAS* 72/2 (2009): 282.

68 *Raṭl* (rotl) was a unit of both weight and capacity.

69 T. Lewicki, "Turkische Stadt Dana nach einem Bericht des arabischen Geographen Ibn al-Faqih", *Folia Orientalia* 13 (1971): 165; *idem*, *Źródła arabskie do dziejów słowiańszczyzny* (Wrocław, 1969), 42.

70 C.E. Bosworth, "Al-Xwarazmi on the People of Central Asia", *JRAS* (1965): 6–12 (reprinted in *idem*, *The Medieval History of Iran, Afghanistan and Central Asia* (London, 1977), art. 20).

71 Baward in Turkmenistan.

72 Nisa in Khorasan.

73 On the Oxus River.

74 Jere L. Bacharach, "The Career of Muhammad ibn Tughj al-Ikhshid, a Tenth Century Governor of Egypt", *Speculum* 50 (1975): 595.
75 Johan Elverskog, *Buddhism and Islam on the Silk Road* (Philadelphia, 2010), 68.
76 Van Laer, "Ibn al-Faqīh et le Livre des Pays", 73–4.
77 Vladimir Minorsky, "La deuxieme risāla d'Abū-Dulaf", *Oriens* 5/1 (1952): 23–7.
78 C. von Schlözer, *Abū Dolef Misaris de itinere Asiatico commentarius* (Berlin, 1845); F. Wüstenfeld, "Des Abū Dulaf Bericht über die türkischen Horden", in *Zeitschrift für vergleichende Erdkunde* Band 2 (1842): 205 ff. (German translation); A. von Rohr-Sauer, *Abū Dulaf Bericht uber seine Reise nach Turkestan* (Bonn, 1933); in Fami Saʿd and Talal Majdhub (eds.), *Taḥqīq al-makhṭūṭāt bayna al-naẓariyya wal-taṭbīq* (Beirut, 1993); Abū Dulaf Misʿar b. Mahalhal al-Yanbūʿī al-Khazrajī (d. 1000), *al-Risāla al-ulá*, ed. Marzin Saʿid Marzin Asiri (Mecca, 1995).
79 Qurʾān 30, al-Rūm (Byzantium): 9; cf. 27, al-Naml (the Ants): 51–2.
80 Qurʾān 45: 37.
81 Luke Treadwell, "The Account of the Samanid Dynasty in Ibn Ẓāfir al-Azdī's *Akhbār al-duwal al-munqaṭiʿa*", *Iran* 43 (2005): 152.
82 J. Marquart, *Osteuropäische und ostasiatische Streifzüge Ethnologische und historisch-topographische Studien zur Geschichte des 9. und 10. Jahrhunderts (ca. 840–940)* (Leipzig, 1903), 74–5.
83 Ghada al-Hijjawi al-Qaddumi (trans.), *Book of Gifts and Rarities: Selections Compiled in the Fifteenth Century from an Eleventh-century Manuscript on Gifts and Treasures* (Cambridge, MA, 1996), 155–64, refers to a Chinese delegation in 327/938–939; C.E. Bosworth, "An alleged embassy from the Emperor of China to the Amir Nasr b. Ahmad: A contribution to Samanid military history", in M. Minovi and I. Afshar (eds.), *Yad Nama-yi irani-yi Minorsky* (Tehran, 1969), 17–29.
84 Consumption of millet served as a criterion of social manners. See al-Kāshgharī/ Dankoff, *Compendium*, 1: 100: "*dukhn* among the Turks, the Oghuz (*ghuzziyya*) do not know this".
85 Minorsky (ed. and trans.), *Marvazi on China, the Turks and India*, Arabic text (*c*. A.D. 1120) with an English translation and commentary (London: The Royal Asiatic Society, 1942), 29, 32–3.
86 D.G. Tor, "'Sovereign and Pious': The Religious Life of the Great Seljuq Sultans", in Christian Lange and Songul Mecit (eds.), *The Seljuqs: Politics, Society, and Culture* (Edinburgh, 2011), 42, 57 (note 18).
87 Yves Bonnefoy (ed.), *Asian Mythologies* (Chicago, 1993), 331–2.
88 Describing the major rivers of central Asia, al-Masʿūdī mentions the Irtysh: "On its banks are the summer and winter encampment of the Turk Kimakiyya and Ghuzz tribes." Marquart, *Osteuropäische und ostasiatische Streifzüge*, 79, says that Abū Dulaf is wrong in locating the Kīmāk; Bosworth, "Al-Xwarazmi", 4–5.
89 Michael R. Drompp, "Breaking the Orkhon Tradition: Kirghiz Adherence to the Yenisei Region after A.D. 840", *JAOS* 119/3 (1999): 390–403.
90 Yedida Kalfon Stillman and Norman A. Stillman, *Arab Dress: A Short History from the Dawn of Islam to Modern Times* (Leiden, 2007), 12.
91 Zakarīyā b. Muḥammad al-Qazwīnī (d. 682/1283), *Athār al-bilād wa-akhbār al-ʿibād* (Beirut, 1380/1960), 585: "The Khutlukh are Turkic people. To cross their country takes ten days. They are the most violent Turkic people and frequently attack their neighbours. They are wise and capable. The marriage of sisters is a common practice among them. A woman marries only once in her life. Widowed she will not remarry. Fornicators are punished harshly; the man and the woman are burned to death. Divorce is not practiced among them. The bride price is the whole property of the husband. They consume barley, sweetpeas and wheat as well as uncooked meat. If a man without capital marries a woman he will pay her bride price by serving her guardian for a year. They permit revenge for bloodshed. A wounded person would be compensated for their injuries. Yet if the wounded dies after the collection of payment

his blood will not be avenged. Their king determinedly objects to evil and agrees not with malevolence. If their king decides that a person will not marry and this person does so, he will face execution."

92 Ibid., 583: "They are Turkic people. It takes twenty days to traverse their country. They are intelligent and gifted people, in contradiction to other Turkish tribes"; on this nomadic people, see M. Ed. Chavannes, "Voyageurs Chinois chez les Khitan et les Joutchen", *Journal Asiatique* (1897): 382; Michal Biran, *The Empire of Khitai in Eurasian History: Between China and the Islamic World* (Cambridge, 2005), 33–5; J. Daniel Rogers, "Inner Asian States and Empires: Theories and Synthesis", *Journal of Archaeological Research* 20 (2012): 227–9.

93 al-Qazwīnī, *Athār al-bilād wa-akhbār al-ʿibād*, 606: "A land close to China."

94 This narrative had widespread circulation among ʿAbbāsid-period Arab/Muslim writers. al-Masʿūdi, *Murūj* (Paris, 1863), 1: 350, 3: 154, reports that "the kingdom of Tibet is distinct from the kingdom of China. The majority of the population is Himyarite who governs the land. Several descendants of the royal Yemenite Tubbaʿ dwell amongst them."

95 An ancient pre-Islamic south Arabian kingdom that fell in the sixth century.

96 Silverstein, *Postal Systems*, 93.

97 Ulrich Haarmann, *Quellenstudien zur frühen Mamlukenzeit* (Freiburg, 1970), 61 ff.; Peter Thorau, "Zur Geschichte der Mamluken und ihrer Erforschung", *Die Welt des Orients* 20/21 (1989/1990): 235–36; Howayda al-Harithy, "The Patronage of al-Nāṣir Muḥammad ibn Qalāwūn, 1310–1341", *MSR* 4 (2000): 228.

98 Ibn al-Dawādārī, *Kanz al-Durar wa-jāmiʿ al-ghurar* vol. 7: *al-Durr al-maṭlūb fī akhbār mulūk banī ayyūb*, ed. S. ʿAbd al-Fattāḥ ʿAshūr (Cairo, 1391/1972), 217–24, 226–7; *Durar al-Tijān* in Graf, *Die Epitome der Universalchronik Ibn ad-Dawadaris*, 56–63. (Arabic). In *Kanz al-Durar* the account is inserted in the event of AH 718/1318, where he quotes Ibn al-Athīr's narration of the coming of the Tatars (AH 617/1220–1221). Richards, *The Annals of the Saljūq Turks*, 292–3; in *Durar al-Tijān* it is dated AH 719/1319. The additions in square parenthesis are from *Durar al-Tijān.*

99 Ulrich Haarmann, "Turkish Legends in the Popular Historiography of Medieval Islam", in Frithiof Rundgren (ed.), *Proceedings of the VIth Congress of Arabic and Islamic Studies Visby and Stockholm 1972* (Stockholm/Leiden, 1975), 99–100, 102–3.

100 Ulrich Haarmann, "Altun Han und Cingiz Han bei den aegyptischen Mamluken", *Der Islam* 51/1 (1974): 16–23; Mehmed Fuad Köprülü, *Early Mystics in Turkish Literature*, 18.

101 P.M. Holt, "The Sultanate of al-Mansur Lachin (696–8/1296–9)", *BSOAS* 36 (1973): 525.

102 Jacqueline Sublet, "Nom écrit, nom dit: Les personnages du théâtre d'ombres d'Ibn Dāniyāl", *Arabica* 44/4 (1997): 545–52; Li Guo, "Paradise Lost: Ibn Daniyal's Response to Sultan Baybars' Campaign against Vice in Cairo", *JAOS* 121/2 (2001): 219, 230.

103 Margaret Larkin, "Popular Poetry in the Post-classical Period, 1150–1850", in Robert Irwin (ed.), *The New Cambridge History of Islam* vol. 4 *Islamic Cultures and Societies to the End of the Eighteenth Century* (Cambridge, 2010), 205–7.

104 David J. Roxburgh, "On the Transmission and Reconstruction of Arabic Calligraphy: Ibn al-Bawwab and History", *SI* 96 (2003): 39–53.

105 Haarmann, "Altun Han und Cingiz Han", 6–7.

106 Stillman and Stillman, *Arab Dress*, 52, 135, 167.

107 Saleh Said Agha, "Abū Muslim's Conquest of Khurasan: Preliminaries and Strategy in a Confusing Passage of the Akhbār Al-Dawlah Al-ʿAbbāsiyyah", *JAOS* 120/3 (2000): 333–47; Jacob Lassner, "Abū Muslim al-Khurāsānī: The Emergence of a Secret Agent from Khurāsān, Irāq, or Was it Iṣfahān?", *JAOS* 104/1 (1984): 165–75.

108 C.E. Bosworth, "The Heritage of Rulership in Early Islamic Iran and the Search for Dynastic Connections with the Past", *Iranian Studies* 11 (1978): 16; *idem*, "An

Oriental Samuel Pepys? Abū'l-Faḍl Bayhaqī's Memoirs of Court Life in Eastern Iran and Afghanistan, 1030–1041", *JRAS* 14/1 (2004): 13–25.

109 Said to be the vizier of the Sasanid emperor Khosrow I Anōshīravān (r. 531–78). C.E. Bosworth (trans.), *The History of al-Tabari*: vol. 5 *The Sasanids, the Byzantines, the Lakhmids, and Yemen* (Albany, 1999), 294–5 (note 694).

110 David Cook, *Martyrdom in Islam* (Cambridge, 2007), 108.

111 On "Ulu" see M.T. Houtsma, *Ein türkisch-arabisches Glossar* (Leiden, 1894), 54 (the book was written after 1220, presumably during the Mamlūk era or was at least the property of a Mamlūk *amīr*).

112 The text in the *Durar* (59 ll. 15–17) says: "From the lake a river gushes out. It runs parallel to the mountain's slope to the distance of forty full parasangs. At its end two huge cities were built. The cities' names are, in the Mongol language, Kashirmaq and Aydarmaq." See below.

113 Werner Diem, "Untersuchungen zu Technik und Terminologie der arabisch-islamischen Türschlösser", *Der Islam* 50 (1973): 136–7.

114 Literally: "the great moon that is the primal father".

115 "Serpents"? Cf. Carré de Vaux, "Notice sur un calendrier turc", in *A Volume of Oriental Studies presented to G. Browne* (Cambridge, 1922), 107, 108; Rudolf M. Riefstahl, "Vier Syrische Marmorkapitale mit figuralen Darstellungen in der Moschee zu Boz Uyuk", *Der Islam* 20 (1932): 193; Jean-Paul Roux, "Mosquées anatoliennes à décor figuratif sculpté", *Syria* 57/1 (1980): 317, 320.

116 E. Sachau, *The Chronology of Ancient Nations* (Leipzig, 1923), 173.

117 The fish, Pisces (18 February–20 March). This is the twelfth astrological sign while the ram is the first.

118 Emil L. Fackenheim, "The Possibility of the Universe in Al-Farabi, Ibn Sina and Maimonides", *Proceedings of the American Academy for Jewish Research* 16 (1946–47), 53–4; I.R. Netton, *al-Farabi and his School* (London, 1992), 32–3; Thérèse-Anne Druart, "al-Farabi, Emanation, and Metaphysics", in Parviz Morewedge (ed.), *Neoplatonism and Islamic Thought* (Albany, 1992), 127–48.

119 In *Durar al-Tijān*, 62 (ll. 9–12) the author follows another path: "I say I shortened these legendary stories and stuck to Jibrīl's transmission of the ancient Turkish book, which I had already mentioned above, who says after the lines skipped by us: From this ancestor, whom they name *ulu ay ata-ci* (the Great Moon Father) and from his wife, who by the Turks is called *ulu ay ana-ci* (the Great Moon Mother) as she was formed in a similar manner to creation of the man, a great number of people issued."

120 I omitted pp. 224 (l. 13)-226 (l. 17) that deals with man's life expectancy.

121 In *Durar al-Tijān*, 62 (l. 13): "As their numbers grow some looked for help against the others."

122 In *Durar al-Tijān*, 62 (l. 15): "They consented to have the eldest son of the Ulu Ay Ataci (the Great Moon Father) be their king. So they nominated him as their monarch."

123 See Houtsma, *Ein türkisch-arabisches Glossar*, 64 (Küchlein).

124 In *Durar al-Tijān*, 63 (ll. 1–6): "The place became a shrine for them. Each year on the anniversary of that king's death they gather there. Pilgrims from all over that country assemble there. All pray to that golden mummy that contains the king's corpse. For three days they invoke him and humble themselves. Those who fail to come and demonstrate disobedience are brought to the doors of that shrine and are executed there."

125 Broadbridge, "Academic Rivalry and the Patronage System in Fifteenth-Century Egypt", 86–90; Donald P. Little, "A Comparison of al-Maqrīzī and al-ʿAynī as Historians of Contemporary Events", *MSR* 7/2 (2003): 207–9; Nobutaka Nakamachi, "al-ʿAynī's Chronicles as a source for the Baḥrī Mamlūk Period", *Orient* 40 (2005): 140–71.

126 Badr al-Dīn Maḥmūd al-ʿAynī (1360–1451), *al-Sayf al-muhannad fī sīrat al-malik al-muʾayyad*, ed. F.M. ʿA. Shaltūt (Cairo, 1998), 14–15, 19–22; *idem*, *al-Rawḍ al-zāhir fī sīrat al-malik al-ẓāhir* [Tatar], ed. H. Ernst (Cairo, 1962), 3–4.
127 Qurʾān, 11, al-Hud: 25–50; M.A.S. Abdel Haleem, "The Qurʾanic Employment of the Story of Noah", *Journal of Qurʾanic Studies* 8/1 (2006): 38–57.
128 Gabriel Said Reynolds, "A Reflection on Two Qurʾānic Words (Iblīs and Jūdī), with Attention to the Theories of A. Mingana", *JAOS* 124/4 (2004): 683–8.
129 Cf. "The Prophet said: The nation of ʿĀd was destroyed by *al-dabūr*." R.B. Serjeant, "Star-Calendars and an Almanac from South-West Arabia", *Anthropos* 49/3–4 (1954): 451 (l. 9).
130 Genesis 10: 2 and 1 Chronicles 1: 5 "The sons of Japheth; Gomer, and Magog, and Madai, and Javan, and Tubal, and Meshech, and Tiras"; Donald Daniel Leslie, "Japhet in China", *JAOS* 104/3 (1984): 405–6.
131 Claus Schönig, "On Some Unclear, Doubtful and Contradictory Passages in Maḥmūd al-Kāšġari's 'Dīwān Luġāt at-Turk'", *Türk Dilleri Araştırmaları* 14 (2004): 38–46.
132 Abū Isḥāq Ibrāhim b. Muḥammad al-Fārisī al-Iṣṭakhrī (d. 957), *al-Masālik wal-mamālik* (written in 951), ed. De Goeje (Leiden, 1927), 10 (ll. 8–10), 225 (l. 6), 227 (l. 5) describes them as a group living on the western edges of the steppe, close to the Byzantine frontiers. He says: "A group of Turks, named al-Bajanakiyyah, was cut off from their lands of origin and wandered westwards to the territory between the Caspian Sea and Byzantium. At present they are not dwelling in their ancient homeland but rather in a land that they seized and occupied." For more details see Golden, "The Peoples of the South Russian Steppe", 270–5.
133 Louis Bazin, "Notes sur les mots 'Oġuz' et 'Türk'", *Oriens* 6/2 (1953): 316–17.
134 Golden, "The Peoples of the South Russian Steppe", 351, 355–6.
135 (*Sic*) Presumably al-ʿAynī confused the victories of Ṭughril Beg in eastern Iran with the success of Alp Arsalān in Eastern Turkey or that he used an early source uncritically (al-Kāshgharī? see next note).
136 Cf. al-Kāshgharī/Dankoff, *Compendium*, 1: 101–2. The names in brackets are Kāshgharī's reading.
137 The tribe of the Seljuqid family. Claude Cahen, "Le Malik-Nameh et l'histoire des origines Seljukides", *Oriens* 2 (1949): 51.
138 Mehmed Fuad Köprülü, *The Origins of the Ottoman Empire*, trans. Gary Leiser (Albany, 1991), 44.
139 Altan Çetin, "Oghuz Turks in the Account of Mamluk Historian", *Journal of Islamic Studies* 20/3 (2009): 376–82; Mounira Chapoutot-Remadi, "Turkomans in Syria and Circassian Power", *Mediterranean World* 6 (2010): 48.
140 Doufikae-Aerts, *Alexander Magnus Arabicus*, 135–6.
141 Ibrahim Kafesoğlu, "A Propos du nom Türkmen", *Oriens* 11 (1958): 146–50 offers another explanation.
142 Joel L. Kraemer, *Humanism in the Renaissance of Islam: The Cultural Revival During the Buyid Age* (Leiden, 1986), 281; Andrew C.S. Peacock, "Seljuq Legitimacy in Islamic History", in Christian Lange and Songul Mecit (eds.), *The Seljuqs: Politics, Society and Culture* (Edinburgh, 2011), 80.
143 Bilal Orfali, "The Works of Abū Manṣūr al-Thaʿālibī (350–429/961–1039)", *Journal of Arabic Literature* 40 (2009): 276.
144 Abū al-ʿAlāʾ Muḥammad b. Abī al-Qāsim ʿAlī b. Ḥusayn Ibn Ḥassūl, "*Kitāb tafḍīl al-atrāk ʿalá sāʾir al-ajnād*", ed. A. ʿAzzawi, *Belleten* 4 (1940): 26–51 (Arabic).
145 The noun *fitna* occurs thirty times in the Qurʾān. It is a cornerstone in the religious-political vocabulary of Islam. See, for example, Qurʾān 37: 67 "A trial for the evildoers"; Qurʾān 2: 191, 217 "persecution is more grievous/heinous than slaying"; Qurʾān 2: 193 "till there is no persecution and the religion is God's"; and 7: 73: "Unless you do this, there will be persecution in the land and great corruption."
146 A paraphrase of Qurʾān 2: 209: "If you slip after the clear signs have come to you."

147 Cf. Qur'ān, 3: 119: "Perish in your rage."
148 An allusion to Qur'ān 18: 22: "So do not dispute with them."
149 A paraphrase of Qur'ān 4: 135: "[So] follow not caprice."
150 A paraphrase of Qur'ān 52: 12: "Woe on that day unto the deniers who now find pleasures by playing in vain disputes"; and of Qur'ān 74: 45–47: "We used to wade in vain disputes with those who plunged there into and we used to deny the Day of Judgement till the inevitable came unto us."
151 Cf. Qur'ān 15: 59: "We shall deliver the folk of Lot all together."
152 Goldziher, *Muslim Studies*, 1: 23.
153 *'Iṣma* means protection, as in Qur'ān 5: 67: "God will protect thee from men. God guides not the disbelieving people"; and in Qur'ān 4: 175 "And for those who believe in God, and hold fast to Him, He will surely admit them into His mercy and grace, and will guide them to Him on a straight path." Cf. Qur'ān 3: 101: "Whosoever holds fast to God, he is guided to a straight path"; and 3: 103: "And all of you together attain protection in Allah's bond of salvation"; Shahab Ahmed, "Ibn Taymiyyah and the Satanic Verses", *SI* 87 (1998): 71, 75, 86–7.
154 A reference to Qur'ān 5: 67: "Allah will protect thee"; 3: 101: "And in Allah find refuge from danger"; 22: 78: "Hold fast to Allah. He is your protecting Friend."
155 An allusion to Qur'ān 4: 62: "Swearing by Allah that they were seeking naught but harmony and kindness"; 11: 88: "My welfare is only with God; in Him I have put my trust, and to Him I turn penitent."
156 An allusion to Qur'ān 4: 9: "Let them fear Allah and speak justly"; 19: 50: "We gave them Our mercy, and We appointed unto them a lofty tongue of truthfulness."
157 An allusion to *amāna* in Qur'ān 33: 72: "We offered the Trust to heaven and earth and the mountains"; Cf. Louis Gardet, "Les noms et les statuts. Le problème de la foi et des œuvres en Islām", *SI* 5 (1956): 66.
158 Literally: manifested and concealed, as in Qur'ān 31: 20.
159 An allusion to Qur'ān 33: 70: "O believers, fear God, and speak words hitting the mark (*qawl[an] sadīd[an]*), and He will set right your deeds"; and see Frederick S. Carney, "Some Aspects of Islamic Ethics", *The Journal of Religion* 63/2 (1983): 163.
160 Cf. the royal Saljūqid title *muẓhir al-'adl wal-inṣāf*; J.M. Rogers, "Waqf and Patronage in Seljuk Anatolia: The Epigraphic Evidence", *Anatolian Studies* 26 (1976): 70; Ann K.S. Lambton, "Changing Concepts of Justice and Injustice from the 5th/11th Century to the 8th/14th Century in Persia: The Saljūq Empire and the Ilkhanate", *SI* 68 (1988): 33.
161 An allusion to Qur'ān 38: 22: "So judge between us justly, and transgress not."
162 In fulfilment of the command in Qur'ān 6: 152: "And give just measure and full weight (*mizān*)."
163 A paraphrase of Qur'ān 2: 86: "Those who buy the life of the world at the price of the hereafter."
164 A paraphrase of Qur'ān 28:50: "And the person who goes astray without guidance from God, the one who follows his caprice"; and also cf. Qur'ān 42: 15.
165 M. Amir Moezzi, "Ibn 'Aṭā al-Adamī, esquisse d'une biographie historique", *SI* 63 (1986): 89; Gerhard Böwering, "Ideas of Time in Persian Sufism", *Iran* 30 (1992): 78.
166 That is to say, people who act contrary to the teachings of Qur'ān 29: 36: "Serve Allah and look forward to the Last Day, and do not evil."
167 Compare Qur'ān 30: 30: "As a worshiper of the true faith set your face to the religion, God's original religion, upon which He originated mankind." On *fiṭra* as the true religion, that is, Islam, see Gérard Lecomte, "Le problème d'Abū 'Ubayd: Réflexions sur les 'erreurs' que lui attribue Ibn Qutayba", *Arabica* 12/2 (1965): 161. The official Saljūq ideology ascribed considerable importance to their image as protectors of Islam. 'Imād al-Dīn Muḥammad al-Kātib al-Iṣfahānī (1125–1201) composed a chronology of the Saljūq viziers, entitled *Nuṣrat al-fatra wa-'uṣrat al-fiṭra fī akhbār al-wuzarā' al-saljūqiyya* (*The Support of the Period and the Shelter of Islam: The*

History of the Saljūq Bureaucracy). This work was studied recently by David Durand-Guédy, "Memoires d'exiles: lecture de la chronique des Saljuqides de Imad al-Din al-Isfahani", *Studia Iranica* 35 (2006): 181–202; and A.C.S. Peacock, "'Imad al-Din al-Isfahani's *Nusrat al-Fatra*, Seljuq politics and Ayyubid origins", in Robert Hillenbrand, A.C.S. Peacock and Firuza Abdullaeva (eds.), *Ferdowsi, The Mongols and the History of Iran: Art Literature and Culture from Early Islam to Qajar Persia* (London, 2013), 79–91.

168 As was Jonah (the Man of the Fish) condemned. Qur'ān, 68: 49. Cf. also 17: 22.

169 Referring to Qur'ān 33: 38: "And God's commandment is certain destiny."

170 This line paraphrases Qur'ān 2: 256: "There is no compulsion in religion. The right direction is henceforth distinct from error."

171 Referring to Qur'ān 7: 146: "[And] though they see the way of rectitude they will not take it for a way."

172 Qur'ān 71: 14: "What ails you that you look not for majesty in God, when He created you by stages?"

173 On this, see Julian Obermann, "Political Theology in Early Islam: Ḥasan Al-Baṣri's Treatise on Qadar", *JAOS* 55 (1935): 143; J. Meric Pessagno, "The Uses of Evil in Maturidian Thought", *SI* 60 (1984): 70–1.

174 The crown was a symbol of royal glory and omnipotence. The manuscript had been the subject of a heated debate. Muhammad Sabir Khan, "Studies in the Kitāb al-Taǧī of al-Ṣābī", *Arabica* 17 (1970): 151–60; *idem*, "The Contents of the Kitāb al-Tājī Manuscript of Abū Isḥāq Ibrāhīm Aṣ-Ṣābī", *Islamic Studies* 8/3 (1969): 247–52; Wilferd Madelung, "Further Notes on al-Sabi's Kitab al-Taji", *Islamic Studies* 9 (1970): 81–8.

175 'Aḍud al-Dawla ordered al-Ṣābī to compose this book. It narrates the story of Daylam and the Buwayhid dynasty. W. Madelung, "Abū Ishaq al-Ṣābī on the 'Alīds of Tabaristan and Gilan", *JNES* 26 (1967): 17–57. For more details on him see the introduction; Claude Cahen, "L'historiographie arabe: des origines au VIIe s. H.", *Arabica* 33/2 (1986): 151.

176 Cf. Edward G. Browne (trans.), *Abridged Translation of the History of Tabaristan* (compiled in about 613/1216 by Muḥammad b. al-Ḥasan b. Isfandiyār) (London, 1905), 90.

177 The Daylamite people (*Daylamān* in Persian) inhabited the Alburz (Elburz) range and the mountainous region southwest of the Caspian Sea coast. A summary of the medieval Arabic and Persian geographical sources concerning the country of Daylam and Jilan is provided by Le Strange, *The Lands of the Eastern Caliphate*, 172–5.

178 *Dawla* is a knotty political term. The 'Abbāsids first honoured viziers with it, yet the history of this honorific title (*laqab*) is beyond the scope of the present study. It is sufficient to say that the Buwayhids furnished it with a new meaning. They used *dawla* to designate the 'Abbāsid Caliphate and named themselves as its support, prop, staff or arm (*'Imād, Rukn, 'Izz, Mu'izz, 'Aḍud*); Sachau (trans.), *The Chronology of Ancient Nations*, 129; C.E. Bosworth, "The Titulature of the Early Ghaznavids", *Oriens* 15 (1962): 216. For a different interpretation, see W. Madelung, "The Assumption of the Title Shahanshah by the Buyids and the Reign of the Daylam", *JNES* 28 (1969): 96–9.

179 That is to say, the past achievements of the Buwayhid dynasty "have left behind" (Qur'ān, 36: 12) which rewarded them glory and eulogy.

180 Cf. Qur'ān, 39: 63: "His are the keys of the heaven and the earth."

181 H. Amedroz and D S. Margoliouth (trans.), *The Experiences of the Nations* (Oxford, 1921), 6: 311.

182 Can also be translated as enemies; cf. H.F. Amedroz and D.S. Margoliouth (trans.), *The Eclipse of The 'Abbāsid Caliphate: Original Chronicles of the Fourth Islamic Century* (Oxford, 1921), 5: 416, 425.

183 ʿAlī b. Buwayh (Buya; d. 976) was one of three brothers who rose from humble origins to the high position of rulers of western Iran and southern Iraq.
184 Marc Bergé, "al-Tawḥīdī et al-Ǧāḥiẓ: Recensement des textes tawḥīdiens sur la filiation ǧāḥiẓienne d'Abū Ḥayyān al-Tawḥīdī, homme de lettres musulman du IVe/Xe siècle", *Arabica* 12/2 (1965): 188; Franz Rosenthal, "'Blurbs' (*taqrīẓ*) from Fourteenth-Century Egypt", *Oriens* 27 (1981): 177–8.
185 The Hellenistic idea of the "perfect man" (*al-insān al-kamīl*) emerged among Muslims with the development of Islamic neo-Platonism. Majid Fakhry, "The Platonism of Miskawayh and Its Implications for His Ethics", *SI* 42 (1975): 49. The philosopher al-Sijistānī identified the perfect man (*anthropos teleios*) with ʿAḍud al-Dawla. Joel L. Kraemer, "Humanism in the Renaissance of Islam: A Preliminary Study", *JAOS* 104 (1984): 145, 157 (and note 105).
186 A technical term used by Muslim scholars to depict the style and very substance of the sacred books that God has sent down to his messengers. No human being can reproduce such writing. The seeds of this dogma are to be found in the Qurʾān 17: 85–9; 52: 33–5.
187 A play on Qurʾān 7: 97, 98: "Did the people of the village feel secure."
188 Presumably a paraphrase of Qurʾān 51: 54.
189 *akhafa ʿalá* in Qurʾān 7: 59; 11: 3, 26; 26: 135. On the political circumstances, see H.F. Amedroz, "Three Years of Buwaihid Rule in Baghdad, A.H. 389–393", *JRAS* (1901): 503–4.
190 Donohue, *The Buwayhid Dynasty in Iraq*, 270–3.
191 Being the oldest among the second generation of the Buwayhids, Bakhtiyār was allocated the caliph's capital, Baghdad. He was defeated twice by ʿAḍud al-Dawla, who in the second round executed his prisoner in 978; Donohue, *The Buwayhid Dynasty in Iraq*, 51–65.
192 The Arabic *ḥaḍra* (originally: in the presence of) conveys, among other definitions, the double meaning of 'capital city' and 'imperial court'. Armand Abel, "Le Khalife, présence sacrée", *SI* 7 (1957): 32.
193 The ancient Iranian royal epithet was reintroduced by various dynasties that emerged in the lands of the Caliphate. Bosworth, "The Titulature of the Early Ghaznavids", 211; Madelung, "The Assumption of the title Shahanshah by the Buyids", 102.
194 This line echoes Qurʾān 11: 77: "And when Our messengers came to Lot, he was troubled on their account and distressed for them."
195 Probably an allusion to Qurʾān 79: 38: "As for him who was insolent and preferred the present life." Ibn Ḥassūl uses this word as a critical remark on al-Ṣābī's methods.
196 M. Kabir, *The Buwayhid Dynasty of Baghdad* (Calcutta, 1964), 39.
197 The Islamic social code considered the shedding of blood in cases of capital punishment to be additional disparagement, with the purpose of humiliating the accused and his family. The consequences of crucifixion, public exhibition of decapitated bodies, etc. were similar. Ideally, noblemen should be strangled, thus avoiding the shedding of blood. Cf. Ibn Ḥassūl's account of Ibrāhīm Īnāl's execution further below.
198 Baber Johansen, *Contingency in a Sacred Law: Legal and Ethical Norms in the Muslim Fiqh* (Leiden, 1999), 225, 239.
199 The political importance of Qurayshī lineage can be deduced from the supposed saying by Ibn Musayyib: "Had I not been of Quraysh, I would have desired of God that I should be a Persian of Isfahan." Edward G. Browne, "Account of a Rare Manuscript History of Iṣfahán", *JRAS* (1901): 662.
200 Probably a reference to Qurʾān 38: 28: "[Or] shall We make the God-fearing as the transgressors?"
201 This was in line with the ideal virtues of the Muslim ruler. C.E. Bosworth, "An Early Arabic Mirror for Princes: Tahir Dhu al-Yaminain's Epistle to His Son ʿAbd Allah 206/821", *JNES*, 29 (1970): 35.

202 Probably this is a play on words alluding to the Qur'ānic verse (3:155) "God has pardoned them; God is All-forgiving, All-clement (*ḥalīm*)."
203 In the Qur'ān (33: 59), the term *jilbāb* (pl. *jalābīb*) refers to a gown that women wear: "O Prophet: say to thy wives and daughters and the believing women that they draw their cloaks close round them."
204 Probably a reference to Qur'ān, 50: 22: "We have now removed from thee the covering."
205 Literally, those who are familiar with consonants and vowels, or with their shapes and forms. G.W. Thatcher, *Arabic Grammar of the Written Language* (New York, 2003), 5.
206 Literally, "their familiar colours", in opposition to "of diverse colours" in Qur'ān, 35: 27.
207 Allusion to Qur'ān 49: 13: "And We have made you nations (*shu'ūb*) and people."
208 The seat of the Justanids was in Rudbar. Later on they moved to the castle of Alamut, which during the Seljuq era became notorious for being the headquarters of the Assassins. B. Lewis, *The Assassins* (London, 1967), 41–7.
209 The Vahsudan family ruled over Sarjahan and the neighbouring districts during the years that witnessed the disintegration of Buwayhid rule and the expansion of the Ghaznawids. V. Minorsky, "Caucasica IV", *BSOAS* 15/3 (1953): 515–16; W. Madelung, "The Minor Dynasties of Northern Iran", in *Cambridge History of Iran*, 4: 235–6; G.C. Miles, "Numismatics", in *Cambridge History of Iran*, 4: 373
210 The noble offspring of the Prophet's family who were, and still are, respected and admired throughout the Islamic world. Due to their reputation the *ashrāf* wielded political influence in various places and at different times. Hence they were organised into a class headed by a chief [*naqīb*] wearing special clothes, or at least a green turban, and enjoying state allocations.
211 The Muslim Holy Family (*ahl al-bayt*), believed to descend from 'Alī b. Abī Ṭālib and his wife Fāṭima, the daughter of Muḥammad.
212 *Imām* is a complex term. On a daily basis, a Muslim will use it to designate the leader of prayer. In the early Islamic period, as well as in contemporary Iran, it means the legitimate political and religious leader of the Islamic community (*umma*). However, among the Imāmī-Shi'ī denomination, it conveys two particular beliefs. They uphold the belief that the sole legitimate heir of the Prophet Muḥammad is his cousin 'Alī and only he and his progeny are entitled to lead the Muslims. This world vision clearly emerged in the early 'Abbāsid years, during the third quarter of the eighth century, although the party of 'Alī b. Abī Ṭālib was formed much earlier (*c.* 656). The second tenet of this particular branch of Islam is the notion of an absent or hidden imam who will return and re-establish the kingdom of justice on earth. Heinz Halm, *Shiism* (Edinburgh, 1991), 1–155.
213 The cousin and son-in-law of the prophet Muḥammad. Accepted by almost all Islamic sects as the fourth caliph (656–661). His murder by a Khārijī assassin initiated the long chain of martyrs revered by Shi'īs.
214 The Shī'īs and the Sunnīs disagree in their visions of early Islamic history. While the first recognise 'Alī and reject the companions of the Prophet, the latter revere them. Wilferd Madelung, *The Succession to Muhammad: A Study of the Early Caliphate* (Cambridge, 1997), 6–16.
215 Starting with the tenth century, these two schools of law became the dominant interpretation of Islamic jurisprudence in Iraq and Iran. Joseph Schacht, *The Origins of Muhammadan Jurisprudence* (Oxford, 1950/1967), 1–3; N.J. Coulson, *A History of Islamic Law* (Edinburgh, 1964), 50–61, 70–3.
216 al-Ḥasan b. 'Alī al-Aṭrash (the deaf; d. 916), called al-Nāṣir lil-Ḥaqq ("He who seeks to establish God's truth"), defeated the Samanid governor of Tabaristan and converted the non-Muslims in the mountain region of Daylam. His legal interpretation provoked

antagonism between his Nāṣiriyya followers and the Qāsimiyya who founded a Zaydī state in Yemen.

217 Zayd b. ʿAlī Zayn al-ʿAbīdīn b. al-Ḥusayn rebelled against the Umayyads in 739–740. Following the suppression of the revolt, his youngest son, Yaḥyá, fled to Khurāsān. In the ninth century, Zaydiyya scholars in southern Iraq and Medina developed a legal tradition. At the same period a certain Ḥasan b. Zayd established a miniature Zaydī state in northern Iran, supported by Daylamite tribal leaders. In contrast to other Shiʿī branches, the Zaydiyya do not acknowledge a hidden Imam and do not expect the return of the Mahdi. Based upon their particular interpretation of the Qurʾānic verse 33: 33, they assert that all members of the Prophet's family are eligible for the Imamate. Some of the Zaydiyya even recognise the two first heirs of the Prophet, Abū Bakr and ʿUmar, as legal caliphs. Halm, *Shiism*, 206–8.

218 Compare with the data on the royal family of Gilan and Daylam in the second half of the eleventh century that ʿUnṣur al-Maʿālī Kaykavūs b. Iskandar b. Qābūs (1021–1098), who was a contemporary of Ibn Ḥassūl, supplies on his family. Reuben Levy (trans.), *A Mirror for Princes: The Qabus Nama* (London, 1951), 2–3.

219 An Arab tribe mentioned by several pre-Islamic poets who emigrated from Arabia to southern Iraq and Khurāsān.

220 The legend that the people of Daylam are a section of Banū Ḍabba spread in the tenth century, if not earlier. Bosworth, "The Heritage of Rulership in Early Islamic Iran", 13, 18.

221 Al-Ṣābī argued that according to early Arab sources the forefather of the Daylaman originated from the Arab Peninsula.

222 Presumably this use of the verb is a play on Qurʾān 11: 108: "And as for the blessed, they shall be in Paradise."

223 The founding father of the Umayyad dynasty (660–749). R. Stephen Humphreys, *Mu'awiya ibn Abi Sufyan: From Arabia to Empire* (Oxford, 2006).

224 Sunnī Muslims recognise ʿUthmān b. ʿAffān as the third Rightly Guided Caliph (644–655).

225 A war between ʿAlī and his partisans and the old guard of Quraysh was headed by a triumvirate: al-Zubayr, Ṭalḥa and ʿĀʾisha, the widow of the Prophet. Martin Hinds, "The Murder of the Caliph Uthman", *IJMES* 3 (1972): 450–69.

226 Presumably a pun on Qurʾān, 19: 94: "He has indeed counted them, and He has numbered them exactly."

227 For the genealogy of these and other Arab tribes, see Werner Caskel, *Gamharat an-Nasab: das geneologische Werk des Hisham ibn Muhammad al-Kalbi* (Leiden, 1966).

228 According to the Sunnī view, an *imām* must be a scion of Quraysh, the tribe of the Prophet Muḥammad. Bernard Lewis (ed. and trans.), *Islam: From the Prophet Muhammad to the Capture of Constantinople* (New York, 1974), 1: 173; Michel Allard (ed. and trans.), *Textes apologetiques de ʿAbd al-Malik B. ʿAbdallah al-Guwaini (m. 478/1085)* (Beirut, 1968), 177; H.A.R. Gibb, "Al-Mawardi's Theory of the Caliphate", in *idem*, *Studies on the Civilization of Islam* (Boston, 1962), 156; Erwin I.J. Rosenthal, *Political Thought in Medieval Islam: An Introductory Outline* (Cambridge, 1962), 27–9; Ann K. S. Lambton, *State and Government in Medieval Islam: An Introduction to the Study of Islamic Political Theory: The Jurists* (Oxford, 1981), 104–6; Wael B. Hallaq, "Caliphs, Jurists and the Saljūqs in the Political Thought of Juwayni", *The Muslim World* 74 (1984): 34.

229 This title designated a royal prince who served as an executive or a governor. Paul G. Forand, "The Governors of Mosul According to al-Azdī's Ta'rīkh al-mawṣil", *JAOS* 89/1 (1969): 93; Hugh Kennedy, "Central Government and Provincial Elites in the Early ʿAbbāsid caliphate", *BSOAS* 44 (1981): 28.

230 Qurʾān 51: 11.

231 Bahram Gor V reigned during the years 420–438.The nickname *gor* (wild ass) was given to him due to his legendary hunting of an onager. His adventures are a favourite subject in historical legends and manuscript illustrations. Levy, *The Epic of the Kings*, 295ff.; J. Marquart, "Beiträge zur Geschichte und Sage von Erān", *ZDMG* 49 (1895): 660 (refers to al-Ṣābī, *al-Tāji*).

232 Or "the two that almost were slaughtered." Referring to Qur'ān 37: 101–7. Suliman Bashear, "Abraham's Sacrifice of his Son and related Issues", *Der Islam* 67 (1990): 243–77.

233 Cf. the story of Abraham and his son in Qur'ān 37: 99–113.

234 Maxime Rodinson, *Mohammed* (Harmondsworth, 1971), 41–2.

235 A cleaving arrow that abrades the surface of a rock and strikes fire; an allusion to Qur'ān 100: 2.

236 Guillaume (trans.), *The Life of Muhammad*, 66–8.

237 Rosenthal, *Political Thought in Medieval Islam*, 29–30.

238 The three chief Islamic dynasties originated from these clans of Fihr (Quraysh). With the emergence of the Islamic caliphate they were considered the noble relatives of the Prophet Muḥammad (*ahl al-bayt*). Moshe Sharon, "*Ahl al-Bayt*—The People of the House", *JSAI* 8 (1986): 174–80; *idem*, "The Umayyads as *Ahl al-Bayt*", *JSAI* 14 (1991): 135–43.

239 Kinship to the Prophet (*qarāba*) was a significant element taken into consideration during the debate over the question of the rulers' legitimacy. Moshe Sharon, "Notes on the question of the legitimacy of government in Islam", *Israel Oriental Studies* 10 (1980): 121; *idem*, "The Development of the Debate around the Legitimacy of Authority in Early Islam", *JSAI* 5 (1984): 135–6.

240 'Ātika means pure. The traditions of the female ancestry of the Prophet Muḥammad were constructed to reflect the decency of all the mothers in the line, which started with Eve. S. Moinul-Haq (trans.), *Ibn Sa'd's kitab al-Tabaqat al-Kabir* (New Delhi, 1984), 1: 54–63.

241 'Awātik is the collective name of women called 'Ātika. Fawāṭim was the collective designation of a group of women named Fāṭima. The Fāṭimas were two groups of women, either born to the Quraysh tribe or those who married into various clans of that tribe. They were believed to be the mothers of several individuals in the Prophet's lineage.

242 W. Montgomery Watt, *Muhammad at Mecca* (Oxford, 1953), 31.

243 A reference to Qur'ān 4: 146; 31: 22 and 2: 256: "And he who believes in Allah has laid hold of the most firm handle (*al-'urwa al-wuthqa*)."

244 An Arab-Christian warrior and poet (died *c.* 615). He was the son of a black African handmaid. 'Antara b. Shaddād became a famous hero who played a prominent role in medieval Arab popular epics.

245 A water source in eastern Arabia (Yamāma), in the territory of the Banū 'Abs tribe.

246 Peroz, the son of the Sasanian emperor Yazdgard II (fl. 457–84). He spent seven years among the Hephthalites. The Dabuyids, a local dynasty of Tabaristan, claimed to be the descendants of his son, Zamasp/Jamasp (r. 496–98). His figure is incorporated into the Iranian epic. Theodor Noldeke, *The Iranian National Epic or the Shahnamah*, trans. L.T. Bogdanov (Bombay, 1930/Philadelphia, 1979), 16.

247 In this paragraph the author mentions the ancient Persian saga. These stories were remoulded in the form of national epic, as can be deduced from Ferdowsi's *Shah-Nameh*.

248 Muslim scholars worked hard to synchronise Persian mythology with the biblical story of mankind. Mohamad Tavakoli-Targhi, "Contested Memories: Narrative Structures and Allegorical Meanings of Iran's Pre-Islamic History", *Iranian Studies* 29 (1996): 154–6, 158.

249 The king's name is transcribed in several ways: Feredun; Fereydun; Feridun; Faridun. Muslim historians and poets present Afridun as a model hero.

250 Saljūq-period Muslim scholars' vision of world history incorporated these Persian mythological kings in the narrative of past generations. F.R.Ch. Bagley (trans.), [Pseudo] *Ghazali's Book of Counsel for King (nasihat al-muluk)* (London, 1964), 46–54.

251 According to Iranian mythology, with the accession of Zahhak (Ḍaḥḥāk), the arch-evil tyrant in Iranian mythology, evil becomes triumphant everywhere. He reigned one thousand years. See *The Shahnama of Firdausi*, trans. Arthur George and Edmond Warner (London, 1905), 1: 141ff. Eleventh-century Turkic tradition narrates a similar story: Yusuf Khass Hajib, *Wisdom of Royal Glory: (Kutadgu Bilig) A Turko-Islamic Mirror for Princes*, trans. R. Dankoff (Chicago, 1983), 47 (ll. 237–244).

252 Arab authors report that the identity of this mythological hero was a subject of dispute between the Persians and the Arabs. It seems both peoples had adopted him and asserted that he had ruled over them for a period of one thousand years. Ewald Wagner, *Abū Nuwas: eine Studie zur arabischen Literatur der fruhen ʿAbbāsidenzeit* (Wiesbaden, 1965), 137.

253 Cf. the depiction of his historical role in ʿAlāʾ al-Dīn al-Juvaynī, *The History of the World-Conqueror*, trand. J.A. Boyle (Cambridge, MA, 1958), 1: 4–5 (introduction).

254 Levy, *The Epic of the King*, 17–28; Dick Davis (trans.), *The Lion and the Throne: Stories from the Shahnameh of Ferdowsi* (Washington, DC, 1998), 50–81.

255 According to the Iranian division of the universe, the earth (*keshwar*) was divided from North to South into seven zones, each of them split into ten contiguous sections running from East to West. This pre-Islamic division was reported by medieval Muslim geographers. Ibn Khaldun, *The Muqaddimah: An Introduction to History*, trans. Franz Rosenthal (Princeton, 1967), 1: 97.

256 It seems that Central Asian Muslim historians commonly accepted the identification of the mythological Tūr as the ancestor of the Turks. It served them for introducing Eurasian people into the lineage claimed by Iranian and Turkish Muslim dynasties. Barthold, *Turkistan,* 1: 130.

257 Cf. Bosworth, *The History of Al-Tabari*: *The Sasanids, the Byzantines, the Lakhmids, and Yemen*, 302.

258 Levy, *The Epic of the Kings*, 49 (a summary of book 8). In eleventh-century Turkic tradition, Afrasiyab is identified with Tonga Alp Er, an outstanding Turkish prince who conquered Iran. Hajib, *Wisdom of Royal Glory*, 48 (ll. 278–80); Abū al-ʿAlāʾ Aḥval (on him see Dickran Kouymjian, "Problems of Medieval Armenian and Muslim Historiography: The Mxit'ar of Ani Fragment", *IJMES* 4 (1973): 474–5) claims that the Turk Qiniq tribe was the offspring (*az tukhm*) of Afrasiyab; Svat Soucek, *A History of Inner Asia* (Cambridge, 2000), 83; Safi, *The Politics of Knowledge in Pre-modern Islam*, 67.

259 M. Perlman (trans.), *The History of al-Tabarī* 4: *The Ancient Kingdoms* (Albany, 1987), 8ff.; Levy, *The Epic of the Kings*, 99–121; Davis, *The Lion and the Throne*, 247–51 retells Rostam/Rustam's defeat of Afrasiyab.

260 Levy, *The Epic of the Kings*, 81–145; Davis, *The Lion and the Throne*, 117–62.

261 That is to say, the Iranian plateau and Mesopotamia (modern Iraq).

262 *Dhū al-Qarnayn* is the hero of a long passage in the Qurʾān 18: 83–98. The Two-Horned One is generally identified with the great Macedonian king. This interpretation is based among other texts and on the similarity between this excerpt and the *Romance of Alexander*: Babylonian *Talmud*, tractate Middoth, 31–2. However, Muslim sources provide other possible identifications of Dhū al-Qarnayn: Mario Grignaschi, "La Nihayatu-l-ʿArab fi ahbari-l-Furs wa-l-ʿArab (premiere partie)", *BEO* 22 (1969): 18.

263 The image of the famous philosopher as the wise political advisor of the great king was a prevalent subject in Arab and popular Persian literature. Levy, *The Epic of the Kings*, 232–7; Mario Grignaschi, "Le Roman Epistolaire classique conserve dans la version Arabe de Salim Abū al-ʿAla", *Le Muséon* 80 (1967): 221–3; Mahmoud

Manzalaoui, "The Pseudo-Aristotelian '*Kitab Sirr al-asrar*': Facts and Problem", *Oriens* 23 (1974): 163–4.

264 On the duties of the vizier in the Saljūq court, see Lambton, *State and Government in Medieval Islam*, 95–7.

265 George F. Hourani, "Ghazālī on the Ethics of Action", *JAOS* 96/1 (1976): 79.

266 Traditional Zoroastrian historiography expresses deep hatred of Alexander of Macedon. Geoffrey Herman, "Ahasuerus, the Former Stable-Master of Belshazzar, and the Wicked Alexander of Macedon: Two Parallels Between the Babylonian Talmud and Persian Sources", *AJS Review* 29/2 (2005): 295–6.

267 Is it possible that Ibn Ḥassūl refers here to the advice given to Alexander to remove the Persians to Europe? S.M. Stern, *Aristotle on the World-State* (Oxford, 1968), 9 (ll. 20–22), 25.

268 Yet compare with Ḥamza al-Isfahani, *Kitāb ta'rīkh sinī mulūk al-arḍ wal-anbīyā'*, 34: "Alexander had perished and the government of the country fell unto the hands of the kings of the factions (*ṭawâ'if*), they ceased from war and from mutual contention among themselves and one of them would overcome another only by difficult questions. In their days were composed the books which are now in the hands of the people, such as the Book of Maruk, and the Book of Sindibad and the Book of Barsanas and the Book of Shimas and the like, about seventy books in all." D.B. MacDonald, "The Earlier History of the Arabian Nights", *JRAS* (1924): 361.

269 Cf. "Aristotle the teacher of Alexander said: 'The realm must be divided among kings so that they may be occupied with each other and not be engaged with Rum. And they are called the 'party kings'." Kātib/Dabīr Bayhaqī, *Tārīkh-i Mas'ūdī*, 119; Waldman, *Toward a Theory of Historical Narrative*, 154.

270 *Mulūk al-ṭawā'if*, that is, the kingdoms that Alexander's successors set up after his death. Levy, *The Epic of the Kings*, 251.

271 Papak (Babakan 226–240). Levy, *The Epic of the Kings*, 252–4.

272 Starting from the eighth century, this topic captured the minds and imagination of Muslim historians. Bosworth (trans.), *The Sasanids, the Byzantines, the Lakhmids, and Yemen*, 1–21; Pellat (trans.), *Les Prairies d'or*, 1.

273 Cf. Moses' story in the Qur'ān, 10: 88.

274 Qur'ān, 2: 17: "God took away their [the unbelievers] light and left them in darkness unseeing."

275 Gideon Weigert, "A Note on Hudna: Peace Making in Islam", in Y. Lev (ed.), *War and Society in the Eastern Mediterranean, 7–15th Centuries* (Leiden, 1997), 399–406.

276 Lewis, *Race and Slavery in the Middle East*, 45.

277 Hajib, *Wisdom of Royal Glory*, 48 (ll. 276–7).

278 This section of Ibn Ḥaṣṣūl's epistle was studied by David Ayalon, who in order to emphasise his arguments translated selected lines: "The Mamluks of the Seljuks: Islam's Military Might at the Crossroads", *JRAS* (1996): 314–15.

279 Abū Muḥammad 'Abd Allāh b. Muḥammad Ibn al-Sīd (Sayyid) al-Baṭalyawsī (1052–1127), *Kitāb al-ḥadā'iq* (*Il Libro dei Cerchi*), ed. and tr. Massimo Jevolella (Milan, 1984), 135; Max Meyerhof, "An Arabic Compendium of Medico-Philosophical Definitions", *Isis* 10 (1928): 344 (§ 18), 345 (§ 21).

280 The caliphs started to recruit non-Arabs to serve in Islamic armies soon after the Arabs tribesmen reached the frontier zone between Iran and Central Asia. Lewis, *Race and Slavery in the Middle East*, 62–3.

281 Cf. Hajib, *Wisdom of Royal Glory*, 114 (ll. 2270).

282 From their early encounters with the peoples of Central Asia, Byzantine and Arab authors were obsessed with the physiognomy of the Turks. Ibn al-Nafīs, *al-Risāla al-kāmiliyya*, ed. and trans. M. Meyerhof and J. Schacht (Oxford, 1968), 42.

283 In the second half of the fourth century the Roman historian Ammianus Marcellinus described the Huns: "But although they have the form of men, however ugly, they are

so hardy in their mode of life that they have no need for fire or savoury food, but eat the roots of wild plants and the half-raw flesh of any kind of animal whatsoever, which they place between their thighs and on the backs of their horses and thus warm it a little." *Rerum Gestarum*, ed. and trans. J.C. Rolfe (London, 1964), 3: 381–3.

284 Or poverty, as in the Qur'ān 2: 236: "The affluent man according to his means and the needy man according to his means."

285 *Arḍ mujāhil* in the sense of land where people would get lost, as in Qur'ān 6: 35: "If Allah will He could have assembled and brought them all together to the guided road, so be not among the lost."

286 Golden, *An Introduction to the History of the Turkic People*, 212–13.

287 Robert Dankoff, "Kāšġarī on the Beliefs and Superstitions of the Turks", *JAOS* 95/1 (1975): 69.

288 For this ethos, see Hajib, *Wisdom of Royal Glory*, 114 (ll. 2275–81), 116 (2317).

289 Slaves were a common phenomenon in the society of the caliphate. Caliphs considered them loyal servants who could be trusted. A. Nazmi, "The Functions of the White Slaves in the Social Life of the 'Abbasid Caliphate", *Rocznik Orientalistyczny* 51 (1998): 91–103.

290 Turkish soldiers were sent from Baghdad to Egypt as early as the ninth century. The first semi-independent Turkish dynasty to rule Egypt originated from a Turkish officer named Aḥmad b. Ṭūlūn (835-84). See Y. Lev, *State and Society in Fatimid Egypt* (Leiden, 1991), 22, 91.

291 Niẓām al-Mulk, *The Book of Government*, 102–3.

292 Cf. Qur'ān 43: 42: "We have power over them."

293 "*tariku al-turk*[a] *ma tarakukum*." Bashear, *Arabs and Others in Early Islam*, 104–11.

294 Cf. Qur'ān 3: 103: "And hold fast and find security by sticking all of you together in God's covenant and do not split."

295 Bernard Lewis, *Race and Colour in Islam* (New York, 1971), 8–11; *idem*, *Race and Slavery in the Middle East*, 22.

296 Rudolph Peters, *Jihad in Mediaeval and Modern Islam* (Leiden, 1977), 11–25.

297 Everett K. Rowson, "The Effeminates of Early Medina", *JAOS* 111 (1991): 673–91; David Cook, "Moral Apocalyptic in Islam", *SI* 86 (1997): 66.

298 *ishtamala 'alá* means literally to contain, to conceive as in Qur'ān 6: 143, 144: "[Or] what the wombs of the two females contain?"

299 Stefan Leder, *Ibn al-Gauzi und seine Kompilation Wider die Leidenscaft* (Beirut, 1984), 216–17.

300 Passive male homosexuality (*ubna* = sodomy) was considered by the Middle Period Muslims to be an "incurable illness." Franz Rosenthal, "Ar-Razi on the Hidden Illness", *Bulletin of the History of Medicine* 43 (1969): 45–60; Cook, "Moral Apocalyptic", *SI* 86 (1997): 65; Frédéric Lagrange, "L'Obscénité du Vizir", *Arabica* 53/1 (2006): 57, 78.

301 Cf. the proverbs that al-Kashgarī transmits in his eleventh-century dictionary: "When a Turk assumes the morals of a Persian his flesh begins to stink." Robert Dankoff, "Inner Asian Wisdom Traditions in the Pre-Mongol Period", *JAOS* 101/1 (1981): 90–1.

302 For a more balanced evaluation of the Turks, see Ibn Qābūs, *A Mirror for Princes: The Qabous Nameh*, 102–3.

303 A Persian dynasty that ruled in Central Asia (819–1005). R.N. Frye, "The Samanids: A Little-Known Dynasty", *The Muslim World* 34 (1944): 40–5; *idem*, *Bukhara: The Medieval Achievement* (Norman, OK, 1965).

304 George Makdisi, "Les Rapports Entre Calife et Sultan a l'epoque Saljūqide", *IJMES* 6/2 (1975): 229–30.

305 Madelung, "The Assumption of the Title Shahanshah", 107; Geoffrey Khan, "The Historical Development of the Structure of Medieval Arabic Petitions", *BSOAS* 53 (1990): 26–8; Mark R. Cohen and Sasson Somekh, "In the Court of Ya'qub ibn Killis:

A Fragment from the Cairo Genizah", *The Jewish Quarterly Review* 80 (1990): 291 (l. 8); Gelder, "Mirror for Princes or Vizor for Viziers", 318.

306 This title appears on coins struck by Ṭughril Beg in Rayy already in 1048. G. Miles, *The Numismatic History of Rayy* (New York, 1938), 198 (no. 226).

307 The coins of Mas'ūd of Ghazna (r. 1030–40) bear the titles: *nāṣir dīn Allah* and (the keeper of) *ḥāfiẓ 'ibād Allah*. Soheila Amirsoleimani, "Paper and Metal: The Irony That Ensues. The Coinage of Mas'ūd of Ghazna (421–31/1030–40)", *Iran* 40 (2002): 166–7.

308 A gyrfalcon or crested goshawk (*Habichtart*). M.T. Houtsma, *Ein Turkisch-Arabisches Glossar nach der Leidener Handschrift* (Leiden, 1894), 28; Gerhard Doerfer, *Türkische und mongolische Elemente im Neupersischen* (Wiesbaden, 1967), 3: 346–8 (no. 1345). I followed the Arabic transliteration.

309 On biblical names among Turks in the twelfth century, cf. Monika Gronke, "The Arabic Yarkand Documents", *BSOAS* 49 (1986): 492 (doc 3 (l. 2)).

310 Yamīn al-Dawla on a golden *dinar* minted by the sultan Maḥmūd the Ghaznavid (Nishapur 415AH). Stephen Album, *A Checklist of Islamic Coins* (Santa Rosa, 1998), 83a (1606).

311 Literally: God's viceroy on earth. Cf. Patricia Crone and Martin Hinds, *God's Caliphs: Religious Authority in the First Centuries of Islam* (Cambridge, 1986).

312 The author is careful not to use the noun *al-mashrīq wal-maghrīb,* and uses instead the verb form, *sharraqa wa-gharraba*. While the first form implies the Almighty ("To Allah belongs the East and the West", Qur'ān, 2: 115, 142), the second form refers to the old pre-Islamic Turkic system of dividing the government between two co-rulers.

313 This line is an apparent reference to Qur'ān 3: 26: "O God, Master of the Sovereignty, Thou gives the Kingdom to whom You wilt and seizes the Kingdom from whom You wilt."

314 Ann K.S. Lambton, "Justice in the Medieval Persian Theory of Kingship", *SI* 17 (1962): 93; *idem*, "Changing Concepts of Justice and Injustice", 31, 33.

315 A reference to Qur'ān 72: 14: "Those who have converted to Islam ought rectitude."

316 Qur'ān, 51: 10 "Perish the conjectures who are dazed in perplexity."

317 Asma Afsaruddin, *Excellence and Precedence: Medieval Islamic Discourse on Legitimate Leadership* (Leiden, 2002).

318 Miles, *The Numismatic History of Rayy*, 196–7 (no. 223–5).

319 This honorific title refers to Qur'ān, 3: 13: "God confirm with help whom He will."

320 *'Amīd*, namely "the supporter of the Kingdom", was a title given to high-ranking civil servants in the Persian-Islamic governments. It was used by the Buwayhids in Baghdad, the Samanids in Transoxiana and the Ghaznawids in Afghanistan. The Saljūqs, who inherited the administrative institutions of these dynasties, used it when they referred to the sultans' chief advisors.

321 *'ulūm* has two meanings: sciences and religious knowledge. Qur'ān 58: 11; 102: 5.

322 About the essential knowledge that a Muslim is required to learn, see Kai Ka'us, *A Mirror for Princes: The Qabous Nameh*, 147–8, 198.

323 The issue of God's attributes (*ṣifat Allah*) and the fierce theological debate among Muslim scholars concerning God's names go back to the eighth century. W.M. Watt, *Islamic Creeds: A Selection* (Edinburgh, 1994), 41–2, 48.

324 Refers to the traditional reading of Qur'ān 33: 40: "[Muḥammad] is the Messenger of Allah and the Seal of the Prophets." On the development of the dogma of *khātim al-anbiyā'*, see Y. Friedmann, "Finality of Prophethood in Sunni Islam", *JSAI* 7 (1986): 177–215.

325 Cf. "God's command comes" in Qur'ān 16: 1.

326 On the dogma of *i'jāz al-Qur'ān*, see W. Madelung, "The Origin of the Controversy concerning the Creation of the Koran", *Orientalia Hispanica* 1 (1974): 504–25 (reprinted in *idem*, *Religious Schools and Sects* (Variorum, 1985), article 5).

327 A paraphrase of Qur'ān 2: 176 and 3: 3.

328 C.E. Bosworth, "Abū ʿAbdallāh al-Khwārazmī on the Technical Terms of the Secretary's Art", *JESHO* 12/2 (1969): 162–3.
329 This vision of the Arabic language is in line with the arguments that were presented during the *shuʿūbiya* controversy. Jamal F. al-ᶜAttar, "al-Jahiz's View of 'Arabic' in relation to the Holy Qur'an", in *Democracy in the Middle East: Proceedings of the Annual Conference of the British Society for the Middle Eastern Studies* (St. Andrews, 1992), 26.
330 Qurʾān 21: 33, 36: 40: "They float each in an orbit."
331 This saying echoes the verse "We gave Moses the Book, and appointed with him his brother Aaron as minister." Qurʾān 25: 35. See S.D. Goitein, "The Origin of the Vizierate and its True Character", *Islamic Culture* 16 (1942): 255–62, 380–92 (reprinted in Idem, *Studies in Islamic History and Institutions* (Leiden, 1961)); *idem*, "On the Origin of the Term Vizier", *JAOS* 81/4 (1961): 425–6.
332 On this institution, see Carla L. Klausner, *The Seljuk Vezirate: A Study of Civil Administration 1055–1094* (Cambridge, MA, 1973).
333 An Arabicised military term originating from the Persian (*sipah* = cavalry, army; *salar* = head, general), which became prevalent in Middle ʿAbbāsid-period writings.
334 Ibrahim Yināl (Inal; Yenal) a half-brother of Ṭughril Beg who entered into communications with the sultan's rivals and proclaimed open rebellion against him in 1049. Lambton, *State and Government in Medieval Islam*, 103–4; Peacock, *Early Seljūq History*, 67–8.
335 Cf. Psalms 48: 7: "Fear took hold upon them there and pain, as of a woman in travail."
336 Cf. Qurʾān 11: 59 "That was Ad; they denied the signs of their Lord and rebelled against his Messengers."
337 Cf. Qurʾān 40: 19: "He [Allah] knows the treachery of the eyes and what the breast conceals."
338 Read Saljūq. Dunlop, *The History of the Jewish Khazars*, 259.
339 Mirkhwand provides a long description of this episode and quotes the Mūlk-Nāmah (for this spelling see Gregorii Barhebraei (1226–1286), *Chronicon Syriacum* (Paris, 1890), 217 (l. 22) (E. A. Wallis Budge (trans.), *The Chronography of Gregory Abū'l-Faraj Commonly Known as Bar Hebraeus* (Oxford, 1932), 195) as his major source: Muḥammad b. Burhān al-Dīn Khāvandshāh (1433–1498), *Mirchondi historia Seldschukidarum*, ed. J.A. Vullers (Gissen, 1837), 1–2; see Peacock, *Early Seljūq History*, 27. According to Ibn al-Athīr, it was Duqāq Timur Yaligh (the Iron Bow) who struck Yabghu, the ruler of the Turks, and split his head open: Richards, *The Annals of the Saljūq Turks*, 31.
340 Gurbadhakan (modern Gulpaygan], a small town and district in the centre of the Zagros Mountains (Jibal) range, not far away from Hamadhan. Le Strange, *The Lands of the Eastern Caliphate*, 207, 210.
341 A paraphrase on the speech by Joseph, Qurʾān, 12: 46.
342 This motto cannot be traced in the Qurʾān, since such a verse is not found in the Holy Book. However segments of it can be extracted from several verses. 3: 159, 3: 173, 5: 18, 11: 88, 22: 78.
343 A.M. Edde and F. Micheau (trans.), *Al-Makin Ibn al-Amid Chronique des Ayyoubides 602–658/1205–1260* (Paris, 1994), 7–15 (Laleli 1002 should be corrected to 2002).
344 Istanbul, Suleymaniye Library, MS Laleli 2002 (copied in 756/1355–6), ff. 156b–170b.
345 This is al-Makīn's routine style. See Eddé et Micheau (trans.), *Al-Makin Ibn al-Amid*, 17 (note 1).
346 Ibn al-Athīr provides a long section on the history of Duqāq in his account of the events of AH 432. Richards, *The Annals of the Saljūq Turks*, 30–1.
347 Barthold, *Turkestan Down to the Mongol Invasion*, 257/269; C.E. Bosworth, *The New Islamic Dynasties: A Chronological and Genealogical Manual* (Edinburgh, 1996), 181.

348 On this see Sachau (trans.), *The Chronology of Ancient Nations*, 129.
349 Le Strange, *The Lands of the Eastern Caliphate*, 434.
350 Richards, *The Annals of the Saljūq Turks*, 20–21 (AH 433/1041–42), 22 (AH 435/1044).
351 Donohue, *The Buwayhid Dynasty in Iraq*, 27, 32.
352 For a different account of this turmoil, see Richards, *The Annals of the Saljūq Turks*, 93, 99–100.
353 In contemporary Syria. D.S. Rice, "Medieval Harran: Studies on Its Topography and Monuments, I", *Anatolian Studies* 2 (1952): 80.
354 Cf. Richards, *The Annals of the Saljūq Turks*, 102, 106.
355 Compare with "Say: O Allah! Owner of Sovereignty! Thou givest sovereignty unto whom Thou wilt, and Thou withdrawest sovereignty from who Thou wilt." Qur'an, 3: 26 (trans. Pickthall).
356 Cf. Richards, *The Annals of the Saljūq Turks*, 103–4.
357 Ibid., 114–15.
358 Ibid., 118–19.
359 Ibid., 121.
360 G. Le Strange, "Baghdad During the Abbasid Caliphate. A Topographical Summary with a Note of the Contemporary Arabic and Persian Authorities", *JRAS* (1899): 852.
361 Jacob Lassner, "Why Did the Caliph Al-Manṣūr Build Ar-Ruṣafah? A Historical Note", *JNES* 24 (1965): 95–6.
362 Le Strange, "Baghdad during the Abbasid Caliphate", 284.
363 Richards, *The Annals of the Saljūq Turks*, 122.
364 Ibid., 123.
365 Ibid., 124–5.
366 Cf. George Makdisi, "Autograph Diary of an Eleventh Century Historian of Baghdad", *BSOAS* 18 (1956): 249 (note 4).
367 On the western bank of the Tigris, north of Baghdad. G. Le Strange, "Description of Mesopotamia and Baghdād, Written about the Year 900 A.D. by Ibn Serapion. The Arabic Text Edited from a MS. in the British Museum Library, with Translation and Notes", *JRAS* (1895): 38–9; Kenneth Mason, "Notes on the Canal System and Ancient Sites of Babylonia in the Time of Xenophon", *The Geographical Journal* 56/6 (1920): 471.
368 A high-ranking civilian title used by administrators during the Buyid and Saljūqid periods. Amedroz, "Three Years of Buwaihid Rule", 761; Kraemer, "Humanism in the Renaissance of Islam", 242.
369 Richards, *The Annals of the Saljūq Turks*, 125–6.
370 Ibid., 126.
371 Ghada al-Hijjawi al-Qaddumi (trans.), *The Book of Gifts and Rarieties: Selections Compiled in the Fifteenth Century from an Eleventh-century Manuscript on Gifts and Treasures* (Cambridge, MA, 1996), index.
372 A town located on the western bank of the Tigris, a half-way stage between Baghdad and Wasit. Le Strange, *The Lands of the Eastern Caliphate,* 37.
373 The name would be translated as "the well-remembered prince". C.E. Bosworth, "Notes on Some Turkish Names in Abū 'l-Faḍl Bayhaqī's Tārīkh-i Mas'ūdī", *Oriens* 36 (2001): 306.
374 Richards, *The Annals of the Saljūq Turks*, 127.
375 Ibid., 131.
376 The broad mountain region, which the Greeks called Media and the Saljūqs called Iraq 'Ajami (Persian). Le Strange, *The Lands of the Eastern Caliphate*, 185. On coins from Hamadan and Isfahan that bear Ṭughril Beg's name and the year 452, see Sadi S. Kucur, "A Study on the Coins of Ṭughril Beg, the Sultan of the Great Seljuqs", in C. Alfaro, C. Marcos and P. Otero (eds.), *Actas del XIII Congreso Internacional de Numismática* (Madrid, 2003/2005), 2: 1606.

377 Dionisius A. Agius, *Classic Ships of Islam: From Mesopotamia to the Indian Ocean* (Leiden, 2008), 303.
378 Richards, *The Annals of the Saljūq Turks*, 137–8; George Makdisi, "The Marriage of Ṭughril Beg", *IJMES* 1 (1970): 259–75.
379 Richards, *The Annals of the Saljūq Turks*, 139.
380 Ibid., 142–5.
381 Ibn al-Dawādārī (d. 1313), *Kanz al-Durar wa-jāmiʿ al-ghurar* vol. 6: *al-Durra al-muḍiyya fī akhbār al-dawla al-fāṭimiyya*, ed. Ṣalāḥ al-Dīn al-Munajjid (Cairo, 1380/1961), 335 (l. 8), 337 (ll. 11–13).
382 Ibn al-Dawādārī, *Kanz al-Durar wa-jāmiʿ al-ghurar*, vol. 7: *al-Durr al-maṭlūb fī akhbār mulūk bani ayyūb*, 220 (l. 10).
383 Ibn al-Dawādārī, *Kanz al-Durar*, vol. 6: *al-Durra al-muḍiyya fī akhbār al-dawla al-fāṭimiyya*, 347–8.
384 In *Kanz al-Durar*, vol. 7: *al-Durr al-maṭlūb fī akhbār mulūk bani ayyūb*, 217 (ll. 5–6, 13) Ibn al-Dawādārī says: *wa-hakādhā dhakkarahu ayḍ*[an] *Ibn Wāṣil ṣāḥib taʾrīkh baghdād*. I was not able to locate this title amongst Ibn Wāṣil's (1207–1298) chronicle. Two compositions entitled *Taʾrīkh baghdād* are known to me. Certainly Ibn Abī Ṭayfūr's (819–893) history cannot serve as a source on the Saljūqs; Abū Bakr Aḥmad b. ʿAlī al-Khaṭīb al-Baghdādī (1002–1071), *Taʾrīkh baghdād aw madīnat al-salām* might be a source but I have not detected any related item; cf. al-ʿAynī, *al-Sayf al-muhannad fī sīrat al-malik al-muʾayyad*, 171 (l. 7).
385 Literarily the prince of nine (Turkish = *dokuz*) (tribes). On a Ghuzz king called Tokuz Khakan, see Minorsky, *Sharaf al-Zaman Ṭāhir Marvazi on China, the Turks and India*, 29.
386 Sebüktigin of Ghazna reigned 998–1030.
387 Ibn al-Dawādārī, *Kanz al-Durar*, vol. 6: *al-Durra al-muḍiyya fī akhbār al-dawla al-fāṭimiyya*, 336–7.

Bibliography

Primary sources

Arabic

ʿAbd al-Jabbār b. Aḥmad al-Asadābādi (d. 415/1025). *Tathbīt dalāʾil al-nubuwwaʾ*, ed. ʿAbd al-Karīm ʿUthmān (Beirut, 1966).

Abū Dāʾūd, Sulaymān ibn Ashʿath al-Azadi al-Sijistāni (202–275/818–888). *Sunan Abi Dāʾūdi*, ed. Muḥammad M. ʿAbd al-Ḥamīd (Cairo, 1951).

Abū Dulaf Misʿar ibn Muhalhil. *al-Risāla al-ūlā*, ed. Fami Saʿd and Ṭalal Majdhūb (Beirut, 1993); ed. Marzin Saʿīd Marzin al-Asiri (Mecca, 1995).

—. *al-Risala al-thāniya*, ed. V. Minorsky, *Abū Dulaf Misʿar ibn Muhalhil's Travels in Iran (circa A.D. 950)* (Cairo, 1955).

al-Aʿshaʾ. *Dīwān*, ed. M. Husayn (Beirut, 1968).

al-ʿAynī, Badr al-Dīn. *al-Rawḍ al-zāhir fī sīrat al-malik al-ẓāhir* [*Tatar*], ed. H. Ernst (Cairo, 1962).

—. *al-Sayf al-muhannad fī sīrat al-malik al-muʾayyad*, ed. F.M. ʿA. Shaltut (Cairo, 1998).

al-Baghdādī, ʿAbd al-Qāhir b. Ṭāhir (429/1038). *al-Farq bayna al-firaqi*, ed. M.M. ʿAbd al-Ḥamīd (Beirut, 1416/1995).

al-Bakharzī, Abū al-Ḥasan ʿAlī b. al-Ḥasan b. ʿAlī (d. 468/1075–76). *Dumyat al-qaṣr wa-ʿusrat ahl al-ʿaṣr*, ed. ʿAbd al-Fataḥ and M. al-Ḥilū (Cairo, 1968).

al-Balādhurī, Abū al-ʿAbbās Aḥmad b. Yaḥyā b. Jābir (d. *c.* 279/892). *Futūḥ al-Buldān*, ed. M. De Goeje (Leiden, 1866; repr. 1968).

—. *Ansāb al-ashrāf*, ed. S.D. Goitein (Jerusalem, 1936).

al-Balawī, Abū Muḥammad ʿAbd Allah al-Madini (400/1000). *Sirat Aḥmad b. Ṭūlūn*, ed. Muḥammad Kurd ʿAlī (Damascus, 1939).

al-Baqillānī, Abū Bakr Muḥammad b. al-Ṭayyib (338–403/950–1013). *al-Tamhīd fī al-radd ʿala al-mulaḥida al-muʿaṭṭila wal-rāfiḍa wal-khawārij wal-muʿtazila*, ed. M. Al-Khuḍayri and A. Abū Rida (Cairo, 1366/1947).

al-Bīrūnī, Abū al-Rayḥān Muḥammad b. Aḥmad (362–430/973–1048), *al-Athār al-bāqīya ʿan al-qurūn al-khāliyya* [*Chronologie Orientalischer volker*], ed. E. Sachau (Leipzig, 1923) [C. Edward Sachau (trans.), *The Chronology of Ancient Nations: An English Version of the Arabic Text of Athar-Ul-Bakiya of Albiruni, or Vestiges of the Past* (London, 1879)].

al-Bundarī, Qiwām al-Dīn al-Fatḥ b. ʿAlī (586–643/1190–1245). *Tawārīkh aal saljūq* [Histoire des Seldjoucides de l'Iraq], ed. M.T. Houtsma (Leiden, 1889).

al-Gharnaṭī, Abū Ḥāmid Muḥammad, "*Tuḥfat al-albāb wa-nukhbat al-a'jāb*", ed. G. Ferrand, *Journal Asiatique* 207 (1925): 1–148, 193–304; ed. Alī Umar (Port Said, 2003).

al-Hamadānī, Muḥammad b. 'Abd al-Malik. *Takmīlat ta'rīkh al-ṭabarī*, ed. M.A. Ibrahim (Cairo, 1990).

Ḥājji Khalīfa, Muṣṭafā b. 'Abd Allāh Kātib Jalabi [Kātip Çelebi] (1609–1657). *Kitāb kashf al-ẓunūn 'an asāmī al-kutub wal-funūn* (Istanbul, 1941–1943).

al-Ḥusaynī, Ṣadr al-Dīn Abū al-Ḥasan 'Alī (1180–1225). *Akhbār al-dawla al-saljūqiyya*, ed. M. Iqbal (Beirut, 1933; repr. 1984) [Clifford Edmund Bosworth (trans.), *The History of the Seljuq State: a translation with commentary of the 'Akhbār al-dawla al-saljūqiyya'* (London, 2011)].

Ibn Abī Ṭāhir Ṭayfūr, Aḥmad (204–280/819–893). *Kitāb Baghdād*, ed. H. Keller, *Sechster Band des Kitab Baghdad* (Leipzig, 1908).

Ibn al-'Adīm, Kamāl al-Dīn 'Umar ibn Aḥmad Ibn Abī Jarāda (588–660/1192–1262). *Zubdat al-ḥalab min ta'rīkh ḥalab*, ed. S. Dahhan (Damascus, 1954).

—. *Bughyat al-ṭalab fī ta'rīkh ḥalab*, ed. S. Zakkar (Damascus, 1408/1988) [Facsimile edition Frankfurt, 1989].

Ibn 'Asākir. *Ta'rīkh madīnat dimashq* (Damascus 1995).

Ibn al-Athīr, 'Izz al-Dīn 'Alī b. Abī al-Karm Muḥammad al-Jazīrī, *al-Kāmil fī al-ta'rīkh*, ed. al-Duqqaq (Beirut, 1424/2003) [Donald S. Richards (trans.), *The Annals of the Saljūq Turks: Selections from al-Kāmil fil-Ta'rīkh of 'Izz al-Dīn Ibn al-Athīr* (London, 2002)].

Ibn Bābā al-Qāshāni (or al-Qāshī), Abū al-'Abbās Aḥmad b. 'Alī (d. Marv 510/1116–1167). *Rās māl al-nadīm fī tawārīkh a'yān ahl al-islām* [The boon companion's stock in trade], ed. S. Zakkar (Beirut, 1418/1997).

Ibn Buṭlān, Abū al-Ḥasan al-Mukhtār (d. 444 or 454/1052 or 1062). "*Risāla fī shirā' al-raqīq wa-taqlīb al-'abīd*", in 'Abd al-Sallām Hārūn (ed.), *Nawādir al-makhṭūṭāt* (Cairo, 1954).

Ibn al-Dawādārī, Abū Bakr b. 'Abd Allah b. Aybak (d. 713/1313). *Kanz al-Durar wa-jāmi' al-ghurar* vol. 6: *al-Durra al-muḍiyya fī akhbār al-dawla al-fāṭimiyya*, ed. Ṣalāḥ al-Dīn al-Munajjid (Cairo, 1380/1961); vol. 7: *al-Durr al-maṭlūb fī akhbār mulūk banī ayyūb*, ed. Said Abd al-Fattah Ashur (Cairo, 1391/1972).

Ibn Faḍl Allah al-'Umarī, Shihāb al-Dīn Aḥmad ibn Yaḥya (1301–1349). "*Masālik al-Abṣār fī mamālik al-amṣār*", in Klaus Lech (ed.), *Das Mongolische Weltreich: al-Umari's Darstellung der mongolischen Reiche in seinem Werk Masalik al-abṣar fī mamalik al-amṣar* (Wiesbaden 1968).

Ibn Faḍlān, *Risāla*, ed. Zegi Validi Togan, *Ibn Fadlan's Reisebericht* (Leipzig, 1939; repr. 1966), ed. Sami Dahhan, *Risālat Ibn Faḍlān* (Damascus, 1959).

Ibn al-Faqīh al-Hamadhānī, Abū Bakr Aḥmad b. Muḥammad (fl. 902). *Kitāb al-buldān*, abr. M. de Goeje (Leiden, 1885); ed. Y. al-Hadi (Beirut, 1996) [abridgement: Henri Masse (trans.), *Abrégé du livre des pays* (Damascus, 1973)].

Ibn Funduq al-Bayhaqī, Ẓahīr al-Dīn Abū al-Ḥasan 'Alī b. Abī al-Qāsim Zayd (490–565/1105–1169). *Ta'rīkh bayhaq* (Hyderabad, 1968).

Ibn Ḥassūl, Abū al-'Alā' Muḥammad b. Abī al-Qāsim 'Alī b. Ḥusayn. "*Kitāb tafḍīl al-atrāk 'alá sā'ir al-ajnād*", ed. A. 'Azzawi, *Belleten* 4 (1940): 26–51 (Arabic).

Ibn Ḥawqal, Abū al-Qāsim Muḥammad (d. *c.* 367/977), *Kitāb ṣūrat al-arḍ*, ed. J.H. Kramers (Leiden, 1967) [William Ouseley (trans.), *The Oriental Geography of Ebn Haukal, An Arabian Traveller of the Tenth Century* (London, 1800); J.H. Kramers and G. Wiet (trans.), *Ibn Hauqal- Configuration de la Terre* (Paris, 1964)].

Ibn Ḥazm al-Andalusī, Abū Muḥammad ʿAlī b. Saʿīd (384–456/994–1064). *Jamharat ansāb al-ʿarab*, ed. A.M. Harun (Cairo, 1962).

Ibn Hishām. *Sīrat rasūl Allah*, ed. M. al-Saqa (Cairo, 1391/1971) [A. Guillaume (trans.), *The Life of Muhammad* (Oxford, 1955)].

Ibn Iyās, Muḥammad b. Aḥmad al-Ḥanafī (852–930/1448–1524). *Badāʾiʿ al-zuhūr fī waqaʾiʿ al-duhūr* (The Amazing Flowers about the Events of the Times), ed. M. Mustafa [Die Chronik des Ibn Ijas] (Wiesbaden, 1975; repr. Cairo, 2008).

Ibn al-Jawzī, Jalāl al-Dīn Abū al-Faraj ʿAbd al-Raḥmān b. ʿAlī (510–597/1117–1201). *al-Muntaẓam fī tawārīkh al-mulūk wal-umam* (Haydar-Abad, AH1359); ed. S. Zakkar (Beirut, 1995); ed. M. Ata (Beirut, 1992).

—. *Akhbār al-ḥamqā wal-mughaffilīn*, ed. R.J. Raḍwān (Mecca, 1425/2004).

Ibn al-Kalbī, *Jamharat al-nasab*, ed. N. Hasan (Beirut, 1986); ed. W. Caskel, *Gamharat an-Nasab: das geneologische Werk des Hisham ibn Muhammad al-Kalbi* (Leiden, 1966).

Ibn Khaldūn, ʿAbd al-Raḥmān (732–808/1332–1406). *al-ʿIbar: Diwān al-mubtadāʾ wal-khabar*, ed. Kh. Shakhada and S. Zakkar (Beirut, 1420/2000).

—. *The Muqaddimah: An Introduction to History*, trans. Franz Rosenthal (Princeton, 1967).

Ibn Khallikan, Shams al-Dīn (1211–1282). *Biographical Dictionary*, trans. MacGuckin de Slane (London, 1871/New York, 1961).

Ibn Khurdadhbeh [Ibn Khurradādhbih], ʿUbayd Allah ibn ʿAbd Allah (lived *c.* 205–317/820–916). *al-Masālik wal-mamālik*, ed. M. De Goeje (Leiden, 1889; repr.1967).

Ibn al-Makin, Jirjis b. al-ʿAmīd (601–671/1205–1273). *Taʾrīkh al-muslimīn*, MS Istanbul (Laleli, 2002); ed Cl. Cahen, "*Akhbār al-Ayyubiyyin*", *BEO* 15 (1967): 109–84; Oxford MSS, ed. A.B. Hassan (Cairo, 2010).

—. *Historia saracenica, qua res gestae Muslimorum inde a Muhammede arabe*, ed. T. Erpenius (Leiden, 1625).

Ibn Nadīm. *al-Fihrist* (Beirut, n.d.) [B. Dodge (trans.), *The Fihrist of* [Muḥammad ibn Isḥāq] *Ibn al-Nadim* (fl. 987): *A Tenth-century Survey of Muslim Culture* (New York, 1970)].

Ibn al-Nafīs. *al-Risāla al-kāmiliyya*, ed. and trans. M. Meyerhof and J. Schacht (Oxford, 1968).

Ibn al-Naḥḥās al-Dimyāṭī, Abū Zakariyá Aḥmad b. Ibrāhim b. Muḥammad (d. 814/1411). *Tahdhīb kitāb mashariʿ al-ashwaq ilá masariʿ al-ʿushshāq fī faḍāʾil al-jihād*, ed. S.A. al-Khalidi (Beirut, 1419/1999).

Ibn al-Qalānisī (470–555/1077–1160). *Dhayl taʾrīkh dimashq*, ed. H.F. Amedroz (Leiden, 1908); Suhail Zakkar (ed.), *Taʾrīkh dimashq*, (Damascus, 1983) [H.A.R. Gibb (trans.), *The Damascus Chronicle of the Crusades* (London, 1932); Roger Le Tourneau, (trans.), *Damas de 1075 a 1154* (Damascus, 1952)].

Ibn Qutayba. *Al-Maʿārif*, ed. M.I.A. al-Ṣawi (Beirut, 1970).

Ibn Rusteh. *al-Aʿlaq al-nafīsa*, ed. M.J. de Goeje (Leiden, 1861).

Ibn Saʿd. *Kitāb al-Ṭabaqāt al-kabīr*, ed. E. Mittwoch (Leiden, 1905) [trans. S. Moinul-Haq (New Delhi, 1984)].

Ibn al-Wardī, Sirāj al-Dīn (d. 861/1457). *Kharīdat al-ajāʾib wa-farīdat al-gharāʾib*, ed. M. Fakhuri (Aleppo, 1411/1991).

Ibn Wāsil, Jamāl al-Dīn Muḥammad b. Sālim (604–697/1207–1298). *al-Taʾrīkh al-ṣāliḥī*, MS Istanbul, Suleymaniye Library, Fatih 4224 [*al-Taʾrīkh al-ṣāliḥī: sīrat al-nabī wal-anbīyāʾ wal-khulafāʾ wal-mulūk wa-ghayruhum*, ed. U.A. Tadmuri (Sidon, 2010)].

Ibn Ẓāfir al-Azdī, ʿAlī. *Akhbār al-duwal al-Munqaṭiʿa*, ed. A. Ferre (Cairo, 1972).

al-Idrīsī (560/1165). *Nuzhat al-mushtāq fī ikhtirāq al-āfāq* [Opus geographicum] (Naples, 1984; repr. Cairo, 1994).

al-Iṣfahānī, Ḥamza b. al-Ḥasan. *Ta'rīkh sinī mulūk al-arḍ wal-anbiyā'* (Beirut, 1961).

al-Iṣfahānī, 'Imād al-Dīn Muḥammad al-Kātib (1125–1201). *Kitāb zubdat al-nuṣra wa-nukhbat al-'uṣra*, ed. M.T. Houtsma (Leiden, 1889).

al-Jāḥiẓ, "*Risala ila al-Fath b. al-Khaqan fī manaqib al-atrak*", ed. G. Van Vloten, *Tria Opuscula Auctore Abū Othman Abr ibn Bahr al-Djahiz* (Leiden, 1903); ed A.M. Harun, *Rasā'īl al-jāḥiẓ* vol. 1 (Cairo, 1384/1964), ed. R. Sheshen (Istanbul, 2002) [C.T. Harley Walker (trans.), "Jahiz of Basra to al-Fath ibn Khaqan on the Exploits of the Turks and the Army of the Khalifate in general", *JRAS* (1915): 631–97; W.M. Hutchins, *Nine Essays of al-Jahiz* (New York, 1975)].

al-Juzjānī, Minhāj al-Dīn Sirāj al-Dīn Abū 'Umar 'Uthmān. *Tabaqāt-i Nāṣiri ya ta'rīkh-i irān-u islām*, ed. A. Ḥabibi (Tehran, 1343 shamsi).

al-Kāshgharī, Maḥmūd b. al-Ḥusayn b. Muḥammad (fl. 464–476/1072–1094). *Kitāb Dīwān lughāt al-turk,* facsimile volume in Besim Atalay (trans.), *Divanü Lûgat-it-Türk* (Ankara, 1941) [R. Dankoff (trans.), *Compendium of the Turkic Dialects* (Cambridge, MA, 1982–85)].

al-Kindī, Abū 'Umar Muḥammad (283–350/897–961). *Governors of Egypt*, ed. Rhuvon Guest (London, 1912).

al-Kutubī, Muḥammad ibn Shākir (d. 764/1363). *Fawāt al-Wafāyāt*, ed. I. 'Abbas (Beirut, 1974).

Madelung, W. (ed.). *Arabic Texts Concerning the History of the Zaydi Imams of Tabaristan, Daylaman and Gilan* (Beirut, 1987).

al-Maqdisī, al-Muṭahhar b. Ṭāhir (fl. *c.* 350/960). *al-Bad' wal-ta'rīkh* [Le livre de la creation et de l'histoire], ed. Cl. Huart (Paris, 1899–1903).

al-Maqrīzī, Taqī al-Dīn Aḥmad b. 'Alī (766–845/1364–1441). *Kitāb al-Muqaffā al-kabīr*, ed. M. al-Ya'lawi (Beirut, 1411/1991).

al-Mas'ūdī, Abū al-Hasan 'Alī (d. *c.* 345/956). *Kitāb al-tanbīh wal-ishrāf*, ed. M.J. De Goeje (Leiden, 1893) [B. Carra de Vaux (trad.), *Le livre de l'avertissement et de la revision* [par Macoudi] (Paris, 1896)].

—. *Murūj al-dhahab wa-ma'ādin al-jawahir* (Paris, 1863/Beirut, 1964) [Ch. Pellat (trans.) *Les Prairies d'or* (Beirut, 1974); Lunde and Stone (trans.), *The Meadows of Gold* (London, 1989)].

Miskawayhi, Abū 'Alī Aḥmad b. Muḥammad al-Rāzī (320–421/932–1030). *Kitāb tajārib al-umam*, ed. H. Amedroz (Oxford, 1914); ed. Abū al-Qasim Emami (Tehran, 1422/2001); [H. Amedroz and D. S. Margoliouth (trans.), *The Experiences of the Nations: The Eclipse of The 'Abbāsid Caliphate: Original Chronicles of the Fourth Islamic Century* (Oxford, 1921)].

—. *al-Ḥikma al-khālida*, ed. A. Badawi (Beirut, 1980).

al-Muqaddasi, Shams al-Dīn Muḥammad ibn Aḥmad (completed his book 375–378/985–988). *Aḥsan al-taqāsīm fī ma'rifat al-aqālīm*, ed. M.J. De Goeje (Leiden, 1906/1967).

Muqāṭil b. Sulayman. *Tafsīr*, ed. A.M. Shakhata (Cairo, 1984).

al-Mutanabbī, Abū al-Ṭayyib Aḥmad b. al-Ḥusayn (915–965). *Dīwān* with the *sharḥ* (commentary) by Abū al-Ḥasan 'Alī b. Aḥmad al-Wāḥidī al-Naysābūrī, ed. Friedrich Dieterici (Berlin, 1861).

al-Nābigha. *Dīwān al-safirayn*, ed. Y. Sh. Farḥat (Beirut, 1992).

Nu'aym b. Ḥammād (d. 229/843). *Kitāb al-fitan*, ed. M.M. Sayyid al-Shura (Beirut, 1418/1997).

al-Nuwayrī, Shihāb al-Dīn Aḥmad (677–733/1278–1333). *Nihāyat al-arab fī funūn al-adab* (Cairo, 1923).

al-Qalqashandī, Abū al-ʿAbbās Aḥmad b. ʿAlī (756–821/1355–1418). *Subḥ al-aʿshā fī ṣināʿat al-inshāʾ* (Cairo, 1963).

al-Qazwīnī, Zakarīyā b. Muḥammad (d. 682/1283). *Kitāb ʿAjāʾib al-makhlūqāt wa-gharāʾib al-mawjūdāt*, ed. F. Wüstenfeld (Wiesbaden, 1848, repr. 1967).

—. *Athār al-bilād wa-akhbār al-ʿibād* (Beirut, 1380/1960).

al-Qifṭī, Jamāl al-Dīn ʿAlī b. Yūsuf (568–646/1172–1248). *al-Muḥammadūn min al-shuʿarāʾ wa-ashʿaruhum*, ed. H. Maʿammari (Riyad, 1970).

Qudāma ibn Jaʿfar, Abū al-Faraj al-Khaṭīb al-Baghdādī (d. 337/948). *al-Kharāj wa-ṣinaʿat al-kitāba*, ed. M.H. al-Zubaydi (Baghdad, 1981).

al-Qurṭubī, Saʿd ibn Aḥmad al-Andalusī (420–462/1029–1069), *al-Taʿrīf bi-ṭabaqāt al-umam,* ed. G.J. Avval (Tehran, 1977).

al-Ṣābī, Abū Isḥāq Ibrāhīm b. Hilāl (d. 384/994). *al-Muntazaʿ min Kitāb al-tājī fī akhbār al-dawla al-daylamiyya li-Abī Isḥāq al-Ṣābī* [Excerpt from the first part of *Kitāb al-Taji*], ed. M.H. al-Zabaydi (Baghdad, 1977).

—. *al-Mukhtār min rasāʾil abī Isḥāq Ibrāhīm b. Hilāl b. Zahrūn al-Ṣābī*, ed. Sh. Arslan (Beirut, 1966).

al-Ṣābī, Hilāl. *Rusūm dār al-khalīfa*, ed. M. Awwad (Baghdad, 1964).

al-Ṣafadī, Ṣalāḥ al-Dīn Khalīl. *al-Wāfī bil-wafāyāt*, ed. S. Dedering (Damascus, 1959).

al-Sālimī, Nūr al-Dīn ʿAbd Allāh b. Ḥumayyid. *Tuḥfat al-aʿyan bi-sīrat ahl ʿumān* (Cairo, 1931).

al-Samʿānī, Abū Saʿd ʿAbd al-Karim. *Ansāb*, ed. A.A. al-Barudi (Beirut 1408/1988).

Sibṭ Ibn al-Jawzī, Shams al-Dīn Yūsuf (582–654/1186–1257). *Mirʾāt al-zamān fī taʾrīkh al-aʿyān*, ed. Alī Sewim (Ankara, 1968).

al-Simnānī, Abū al-Qāsim ʿAlī b. Muḥammad b. Aḥmad al-Raḥbī (d. 1106). *Rawḍat al-quḍāʾ wa-tarīq al-najāʾ*, ed. Salah al-Din al-Nahi (Amman, 1984).

al-Ṣūlī, Abū Bakr Muḥammad b. Yaḥya (d. 335/947). *Akhbār al-rāḍī wal-muttaqī* from the *Kitāb al-awrāk*, ed. J. Heyworth-Dunne (Cairo, 1936; repr. Beirut, 1982).

al-Ṭabarī, Abū Jaʿfar Muḥammad b. Jarīr (224–310/838–923). *Taʾrīkh al-rusūl wal-mulūk*, ed. M.J. De Goeje (Leiden, 1879/repr. 1964). ed. M.A. Ibrahim (Cairo, 1386/1967).

—. *The History of al-Tabarī* vol. 4: *The Ancient Kingdoms*, trans. M. Perlman (Albany, NY, 1987).

—. *The History of al-Tabari:* vol. 32 *The Reunification of the ʿAbbāsid Caliphate*, trans. C.E. Bosworth (Albany, NY, 1987).

—. *Jāmiʿ al-bayān ʿan taʾwīl al-Qurʾān* (Beirut, 1995).

—. *The History of al-Tabari* vol. 5: *The Sasanids, the Byzantines, the Lakhmids, and Yemen*, trans. C.E. Bosworth (New York, 1999).

al-Tamīmī, Muḥammad b. Aḥmad b. Tamīm (333/944). *Kitāb al-miḥan*, ed. Y. al-Jaburi (Beirut, 1403/1983).

al-Tanūkhī, Abū ʿAlī al-Muḥassin b. ʿAlī (327–384/939–995). *Nishwār al-muḥāḍara wa-akhbār al-mudhākara*, ed. A. al-Shaliji (Beirut, 1971–73).

—. *Kitāb al-Faraj baʿd[a] al-shidda* [The Deliverance after hardship], ed. ʿA al-Shaliji (Beirut, 1987).

al-Tawḥidī, Abū Ḥayyān ʿAlī b. Muḥammad (d. 1023). *Kitāb al-Imtāʿ wal-muʾānasa*, ed. A. Amin and A. al-Zayn (Beirut, 1960).

al-Thaʿālibī, ʿAbd al-Malik b. Muḥammad b. Ismāʿīl (961–1038). *Histoire des rois des Perses*, trans. H. Zotenberg (Paris, 1900).

al-Tha'alibī al-Naysabūrī, Abū Manṣūr 'Abd Allah b. Muḥammad (350–429/961–1038). *Yatīmat al-dahr fī shu'arā' ahl al-'aṣr* (Cairo, 1885).

—. *Kitāb tatimmat al-yatima*, ed. 'Abbas Iqbal (Tehran, 1353/1934–35).

—. *Laṭā'if al-Ma'ārif* (Cairo, 1960) [C. E. Bosworth (trans.), *The Lata'if al-ma'arif of Tha'alibi* (Edinburgh, 1968)].

—. *Tuḥfat al-wuzara'*, ed. H. A. al-Rawi and B. M. al-Safar (Baghdad, 1977); ed. S. Abū Diya (Amman, 1994).

al-'Utbī, Abū Naṣr Muḥammad b. 'Abd al-Jabbār (d. 413–431?). *al-Yamīnī fī sharḥ akhbār al-sulṭān yamīn al-dawla wa-amīn al-milla maḥmūd al-ghaznawī* [387–421], ed. I. Dh. Al-Thamiri. Beirut, 1424/2004.

al-Ya'qūbī, Aḥmad b. Wādiḥ (d. 284/897). *Ta'rīkh* (Historiae), ed. M.T. Houtsma (Leiden, 1883).

—. *Kitāb al-buldān*, ed. De Goeje (Leiden, 1892, BGA 7) [G. Wiet (trans.), *Ya'kubi – Les Pays* (Cairo, 1937)].

Yāqūt al-Ḥamāwī al-Rūmī. *Mu'jam al-Buldān*, ed. G. Wustenfeld (Leipzig, 1869).

Persian

Anonymous. *Ta'rīkh-i sistān* (Tehran, 1935) [Milton Gold (trans.), *The Tārikh-e Sistān* (Rome, 1976)].

Anonymous. *Mujmal al-tavārīkh wal-qiṣaṣ*, Mahmud Omid Salar (ed.), facsimile of the Berlin manuscript (Teheran, 2001); ed. M.T. Bahar (repr. Teheran, 2003).

Anonymous (fl. *c.* 372/982). *Ḥudūd al-'ālam min al-mashriq ilá al-maghrib* (Tehran, 1352/1933) [V. Minorsky (trans.), *The Regions of the World: A Persian Geography* (London, 1970)].

al-Bayhaqī, Abū al-Fazl Muḥammad b. Ḥusayn Kātib/Dabīr (385–470/996–1077). *Ta'rīkh-i mas'ūdi, ma'rūf bih Tārīkh-i bayhaqi*, ed. M.G. Ya Haqqi (Mashhad, 1383/2002); ed. Sa'id Nafisi (Tehran, 1319, 1326 shamsi).

Firdowsi, Abū al-Qasem. *Shāh-Nāmah (The Book of Kings)*, ed. J. Khaleghi-Moṭlagh (New York, 1988) [R. Levy (trans.), *Ferdowsi, The Epic of the Kings* (London, 1967); Dick Davis (trans.), *The Lion and the Throne: Stories from the Shahnameh of Ferdowsi* (Washington DC, 1998)].

Gardīzī (Gerdizi) Abū Asad 'Abd al-Ḥayy b. Ḍaḥḥāk b. Maḥmūd (d. 444/1053). *Zayn al-akhbār* (Tehran, 1968).

al-Ghazālī, Abū Ḥāmid. *Naṣīḥat al-mulūk*, trans. R. Bagley (Oxford, 1964), 45, 46, 71, 106.

Ibn Isfandiyār, Bahā' al-Dīn Muḥammad. *Ta'rīkh-i Tabaristan*, ed. 'Abbās Iqbāl (Tehran 1350 shamsi) [E.G. Browne (trans.), *An Abridged Translation of the History of Tabaristan* (London, 1905)].

Ibn al-Balkhī. *Fars Name*, ed. R. Nicholson and G. Le Strange (London, 1921).

Ibn Funduq al-Bayhaqī, Ẓahīr al-Dīn Abū al-Ḥasan 'Alī b. Abī al-Qāsim Zayd (490–565/1105–1169). *Ta'rīkh-i bayhaq* (Haydar Abad, 1968).

al-Juvaynī, 'Alā' al-Dīn. *The History of the World-Conqueror*, trans. J.A. Boyle (Cambridge, MA, 1958).

Kai Ka'ūs Ibn Iskandar Ibn Qābūs (485/1072). *Kitāb Naṣīḥat nāma ma'rūf beh-qābūs nāma*, ed. R. Levy (London, 1951); *A Mirror for Princes: The Qabous Nameh* trans. R. Levy (London, 1951).

al-Mustawfī al-Qazwīnī, Ḥamd Allah b. Abū Bakr (1281–1350). *Ta'rīkh-i Guzida*, ed. A. Nava'i (Tehran, 1342 shamsi).

Mir Khvand, Ghiyāth al-Dīn b. Humām al-Dīn (1433–1498). *Historia Seldschukidarum*, ed. Johann August Vullers (Gissen, 1837) [J.A. Vullers (trans.), *Geschichte der Seldschuken aus dem Persischen* (Gissen, 1838)].

Mir Khvand, Ghiyāth al-Dīn b. Humām al-Dīn (1433–1498). *Dastūr al-vuzarā': shāmil-i aḥvāl-i vuzarā-yi Islām tā inqirāz-i Taymūrīyān, 914* (Tehran, 1938).

Narshakhī, Muḥammad (d. 332/943). *Ta'rīkh-i Bukhara*, ed. Ch. Schefer (Paris 1893) [R.N. Frye (trans.), *The History of Bukhara* (Cambridge, MA, 1954)].

Niẓām al-Mulk. *Siyar al-mulūk (Siyasat nameh)* (Tehran, 1340) [H. Darke (trans.), *The Book of Government or Rules for Kings* (New Haven, CT, 1960)].

Rashīd al-Dīn, Faẓl Allah, *Jāmi' al-tavārīkh* [*cami al-tavarih*], ed. Ateş (Ankara, 1960).

al-Ravandī, Muḥammad b. 'Alī b. Sulaymān. *Rāḥat al-ṣudūr wa-āyat al-surur* [being the history of the Saljūqs], ed. Muḥammad Iqbal (London, 1921).

Ẓahīr al-Dīn Nīshāpūri. *The History of the Seljuq Turks: From the Jami al-Tawārīkh: an Ilkhanid adaption of Saljūq Nama*, trans. Kenneth Allin Luther and ed. C. Edmund Bosworth (Richmond, Surrey, 2001).

Zarkūb al-Shīrāzī, Abū al-'Abbās Aḥmad (d. 1387). *Shirāz nāmah*, ed. Isma'il Va'iz Javadi (Tehran, 1350/1971).

Syriac

Bar 'Ibraya, Yohanan [Bar Hebraeus Gregarious] (1226–1286), *Makhtevayuta zubana min risha de-breshita 'adma le-shnat 1285* [*Chronicon Syriacum*] (repr. Glane, The Netherlands, 1987) [E. Wallis Budge, *The Chronography of Gregory Abū al-Faraj Bar Hebraeus* (Oxford, 1932)].

Callisthenes, Pseudo-. *The History of Alexander the Great*, ed. and trans. E.A. Wallis Budge (Cambridge, 1889).

Chabot, J.B. (ed.). *Chronique de Denys de Tell-Mahre (maktbonuta demari deyonosiyus tellmarhya)* (Paris, 1895).

— (ed.). *Chronicon pseudo-Dionysianum* (Louvain, 1952).

Harrak, Amir (trans.). *The Chronicle of Zuqnin: part 3&4 A.D. 488–775* (Toronto, 1999).

Wallis Budge, E.A. (trans.) *The Alexander Book in Ethiopia* (London, 1933).

Research and studies

Abbott, Nabia. "Two Buyid Coins in the Oriental Institute", *American Journal of Semitic Languages and Literatures* 56 (1930): 350–64.

Abel, Armand. *Le Roman d'Alexander – Legendaire medieval* (Brussels, 1955).

Afsaruddin, Asma. *Excellence and Precedence: Medieval Islamic Discourse on Legitimate Leadership* (Leiden, 2002).

Agha, Saleh Said. "Abū Muslim's Conquest of Khurasan: Preliminaries and Strategy in a Confusing Passage of the Akhbār Al-Dawlah Al-'Abbāsiyyah", *JAOS* 120/3 (2000): 333–47.

Agius, Dionisius A. *Classic Ships of Islam: From Mesopotamia to the Indian Ocean* (Leiden, 2008).

Ahmed, Shahab. "Ibn Taymiyyah and the Satanic Verses", *SI* 87 (1998): 67–124.

Akhmedov, A. "Astronomy, Astrology, Observatories and Calendars", in M.S. Asimov and C.E. Bosworth (eds.), *History of Civilizations of Central Asia: Age of*

Achievement: A.D. 750 to the End of the Fifteenth Century (Paris, 2000), 4/2: 195–204.

Album, Stephen. *A Checklist of Islamic Coins* (Santa Rosa, 1998).

Alexander, D.G. "Two Aspects of Islamic Arms and Armor", *Metropolitan Museum Journal* 18 (1983): 97–109.

Allard, Michel (ed. and trans.). *Textes apologetiques de ʿAbd al-Malik B. ʿAbdallah al-Guwaini (m. 478/1085)* (Beirut, 1968).

Amabe Fukuzo. *The Emergence of the Abbasid Autocracy: The Abbasid Army, Khurasan and Adharbayjan* (Kyoto, 1995).

Amedroz, H.F. "Three Years of Buwaihid Rule in Baghdad, A.H. 389–393", *JRAS* (1901): 501–36, 749–86.

al-ʿAmīlī, Muḥsin al-Amīn. *Aʿyan al-shiʿa* (Beirut, 1986).

Amir Moezzi, M. "Ibn ʿAṭā al-Adamī, esquisse d'une biographie historique", *SI* 63 (1986): 63–127.

Amirsoleimani, Soheila. "Paper and Metal: The Irony That Ensues. The Coinage of Masʿūd of Ghazna (421–31/1030–40)", *Iran* 40 (2002): 163–71.

Anderson, A.R. *Alexander's Gate, Gog and Magog, and the Enclosed Nations* (Cambridge, MA, 1932).

Asimov M.S. and C.E. Bosworth (eds.). *History of Civilizations of Central Asia: Age of Achievement: A.D. 750 to the End of the Fifteenth Century* (Paris, 1998–2000).

al-ᶜAttar, Jamal F. "al-Jahiz's View of 'Arabic' in Relation to the Holy Qur'an", in *Democracy in the Middle East: Proceedings of the Annual Conference of the British Society for the Middle Eastern Studies* (St Andrews, 1992), 20–33.

Avishur, Y. *A Medieval Translation of the Latter Prophets into Iraqi and Syrian Judaeo-Arabic* (Tel-Aviv, 2000).

Ayalon, David. "The European-Asiatic Steppe: A Major Reservoir of Power for the Islamic World", in *The Proceedings of the 25th Congress of Orientalists, Moscow 1960* (Moscow, 1963), vol. II, 47–52 [reprinted in his *The Mamlūk Military Society: Collected Studies* (Variorum, 1979), art. 8].

—. "Preliminary Remarks on the Mamlūk Military Institution in Islam", in V.J. Parry and M.E. Yapp (eds.), *War, Technology and Society in the Middle East* (Oxford, 1975), 53–5.

—. "Mamlūkīyāt", *JSAI* 2 (1980): 335–7.

—. "The Military Reforms of Caliph al-Muʿtaṣim, Their Background and Consequences", in his *Islam and the Abode of War: Military Slaves and Islamic Adversaries* (Aldershot: Variorum, 1994), 1–39.

—. "The Mamluks of the Seljuks: Islam's Military Might at the Crossroads", *JRAS* (1996): 305–33.

Bacharach, Jere L. "The Career of Muhammad ibn Tughj al-Ikhshid, a Tenth Century Governor of Egypt", *Speculum* 50 (1975): 586–612.

Bahnassi, Afif. "Fabrication des épées de Damas", *Syria* 53 (1976): 281–94.

Bailey, H.W. "To the Zamasp Namak", *BSOAS* 6 (1930): 55–85, 581–60.

Barthold (Bart'old), W. *Turkestan Down to the Mongol Invasion,* 3rd edn (London, 1968/1977).

Bashear, Suliman. *Arabs and Others in Early Islam* (Princeton, 1997).

Bazin, Louis. "Notes sur les mots 'Oġuz' et 'Türk'", *Oriens* 6/2 (1953): 315–18.

Bergé, Marc. "al-Tawḥīdī et al-Ǧāḥiẓ: Recensement des textes tawḥīdiens sur la filiation ǧāḥiẓienne d'Abū Ḥayyān al-Tawḥīdī, homme de lettres musulman du IVe/Xe siècle", *Arabica* 12/2 (1965): 188–95.

al-Bili, 'Uthman Sayyid Aḥmad Isma'il. *Prelude to the Generals: A Study of Some Aspects of the Reign of the Eighth 'Abbasid Caliph, Al-Mu'tasim Bi-Allah (218–277 AH /833–842 AD)* (Reading, 2001).

Blachère, R. *Abū Ṭayyib al-Motanabbi* (Paris, 1935).

Böwering, Gerhard. "Ideas of Time in Persian Sufism", *Iran* 30 (1992): 77–89.

Bonner, Michael. "The Naming of the Frontier: 'Awasim, Thughur, and the Arab Geographers", *BSOAS* 57/1 (1994): 17–24.

Bosworth, C.E. "The Titulature of the Early Ghaznavids", *Oriens* 15 (1962): 210–33.

—. *The Ghaznavids: Their Empire in Afghanistan and Eastern Iran 994–1040* (Edinburgh, 1963).

—. "Al-Xwarazmi on the People of Central Asia", *JRAS* (1965): 6–12 [repr. in his *The Medieval History of Iran, Afghanistan and Central Asia* (London 1977) art. 20].

—. "Abū 'Abdallāh al-Khwārazmī on the Technical Terms of the Secretary's Art", *JESHO* 12/2 (1969): 113–64.

—. "An Alleged Embassy from the Emperor of China to the Amir Nasr b. Ahmad: A Contribution to Samanid Military History", in M. Minovi and I. Afshar (eds.), *Yad Nama-yi irani-yi Minorsky* (Tehran, 1969), 17–29.

—. "An Early Arabic Mirror for Princes: Tahir Dhu al-Yaminain's Epistle to his Son 'Abd Allah 206/821", *JNES*, 29 (1970): 25–41.

—. "Barbarian Incursion: The Coming of the Turks into the Islamic world", in D.S. Richards (ed.), *Islamic Civilisation 950-1150* (Oxford, 1973), 1–16.

—. *The Later Ghaznavids: Splendour and Decay 1040–1186* (Edinburgh, 1977).

—. "The Heritage of Rulership in Early Islamic Iran and the Search for Dynastic Connections with the Past", *Iranian Studies* 11 (1978): 7–34.

—. "Madyan Shu'ayb in Pre-Islamic and Early Islamic Lore and History", *Journal of Semitic Studies* 29 (1984): 53–64.

—. *The New Islamic Dynasties: A Chronological and Genealogical Manual* (Edinburgh, 1996).

—. "Notes on Some Turkish Names in Abū 'l-Faḍl Bayhaqi's Tārīkh-i Mas'ūdi", *Oriens* 36 (2001): 299–313.

—. "An Oriental Samuel Pepys? Abū'l-Faḍl Bayhaqī's Memoirs of Court Life in Eastern Iran and Afghanistan, 1030–1041", *JRAS* 14/1 (2004): 13–25.

Boyle, John A. "Turkish and Mongol Shamanism in the Middle Ages", *Folklore* 83 (1972): 177–93.

—. "The Alexander Legend in Central Asia", *Folklore* 85 (1974): 217–28.

—. "The Alexander Romance in the East and West", *Bulletin of the John Rylands University Library of Manchester* 60 (1977): 13–27.

—. "Alexander and the Mongols", *Journal of the Royal Asiatic Society* 2 (1979): 123–36.

Brand, Charles M. "The Turkish Element in Byzantium, Eleventh-Twelfth Centuries", *Dumbarton Oaks Papers* 43 (1989): 1–25.

Broadbridge, Anne F. "Academic Rivalry and the Patronage System in Fifteenth-Century Egypt: al-'Ayni, al-Maqrizi and Ibn Hajar al-'Asqalani", *MSR* 3 (1999): 85–107.

Browne, E.G. "Account of a Rare Manuscript History of Iṣfahán", *JRAS* (1901): 411–46.

—. "Account of a Rare, if not Unique, Manuscript History of the Seljuqs", *JRAS* (1902): 567–610, 849–87.

Bulliet, Richard W. "The Shaikh al-Islam and the Evolution of Islamic Society", *Studia Islamica* 35 (1972): 53–67.

—. *Cotton, Climate, and Camels in Early Islamic Iran: A Moment in World History* (New York, 2009).

Burgel, Ch. *Die Hofkorrespondenz Adud al-Dawlas* (Wiesbaden, 1965).

Busse, H. *Chalif und Grosskoning* (Beirut, 1959).

Cahen, Claude. "Le Malik-Nameh et l'histoire des origines Seljukides", *Oriens* 2 (1949): 31–65.

—. "Une corrépondance Buyid inedite", *Studi Orientalistici in onore di G. Levi Della Vida* (Rome, 1956), 1: 83–97.

—. "The Historiography of the Seljuqid Period", in B. Lewis and P.M. Holt (eds.), *Historians of the Middle East* (London, 1962), 59–78.

—. "Al-Makīn ibn al-ʿAmīd et l'historiographie musulmane: un cas d'interpénétration confessionnelle", *Orientalia Hispanica sive studia F. M. Pareja octogenario dicata*, ed. J.M. Barral (Leiden, 1974), 158–61.

—. "L'historiographie arabe: des origines au VIIe s. H.", *Arabica* 33/2 (1986): 133–98.

Canard, Marius. "La relation du voyage d'Ibn Fadlan chez les Bulgares de la Volga", *Annales de l'Institut d'Etudes Orientales de la Faculté des Lettres d'Alger* 16 (1958): 41–146.

Carney, Frederick S. "Some Aspects of Islamic Ethics", *The Journal of Religion* 63/2 (1983): 159–74.

Çetin, Altan. "Oghuz Turks in the Account of Mamluk Historians", *Journal of Islamic Studies* 20/3 (2009): 376–82.

Chapoutot-Remadi, Mounira. "Turkomans in Syria and Circassian Power", *Mediterranean World* 6 (2010): 47–55.

Clauson, Gerald. *An Etymological Dictionary of Pre-Thirteenth-Century Turkish* (Oxford, 1972).

Coello, Pilar. "Las Actividades de las esclavas segun Ibn Butlan (s. 11) y al-Saqati (ss. 12–13) de Malaga", in Maria J. Viguera (ed.), *La Mujer en al-Andalus: Reflejos historicos de su actividad y categories socials* (Madrid, 1989), 201–10.

Cohen, Mark R. and Sasson Somekh. "In the Court of Yaʿqub ibn Killis: A Fragment from the Cairo Genizah", *The Jewish Quarterly Review* 80 (1990): 283–314.

Cook, David. "Moral Apocalyptic in Islam", *SI* 86 (1997): 37–69.

—. *Martyrdom in Islam* (Cambridge, 2007).

Coulson, N.J. *A History of Islamic Law* (Edinburgh, 1964).

Crone, Patricia. *Slaves on Horses: The Evolution of the Islamic Polity* (Cambridge, 1980).

—. "The Significance of Wooden Weapons in al-Mukhtar's revolt and the 'Abbasid revolution", in I. R. Netton (ed.), *Studies in Honour of Clifford Edmond Bosworth* (Leiden, 2000), 1: 174–87.

— and Martin Hinds. *God's Caliphs: Religious Authority in the First Centuries of Islam* (Cambridge, 1986).

Dankoff, Robert. "Kashgari on the Tribal and Kingship Organization of the Turks", *Archivum Ottomanicum*, 4 (1972): 23–42.

—. "Inner Asian Wisdom Traditions in the Pre-Mongol Period", *JAOS* 101/1 (1981): 87–95.

Dickens, Mark. *Medieval Syriac Historians' Perception of the Turks*. MPhil dissertation, University of Cambridge, 2004.

Diem, Werner. "Untersuchungen zu Technik und Terminologie der arabisch-islamischen Türschlösser", *Der Islam* 50 (1973): 98–156.

Donini, G. *Arab Travelers and Geographers* (London, 1991).

Donohue, J.J. "Three Buwayhid Inscriptions", *Arabica* 20 (1973): 74–80.

—. *The Buwayhid Dynasty in Iraq 334 /945 to 403/1012: Shaping Institutions for the Future* (Leiden, 2003).

van Donzel, Emeri, Andrea Schmidt with a contribution by Claudia Ott. *Gog and Magog in Early Eastern Christian Syriac and Islamic Sources: Sallam's Quest for Alexander's Wall* (Leiden, 2010).

Doufikae-Aerts, Faustina. *Alexander Magnus Arabicus: A Survey of the Alexander Tradition Through Seven Centuries from Pseudo-Callisthenes to Suri* (Leuven, 2010).

Dowsett, C.J.F. "A Neglected Passage in the *History of the Caucasian Albanians*", *BSOAS* 19/3 (1957): 456–68.

Dozy, R. *Supplément aux Dictionnaires Arabes* (Leiden, 1881).

Dunlop, M.D. *The History of the Jewish Khazars* (Princeton, 1954).

Eddé, A.-M. and F. Micheau (trans.). *Al-Makin Ibn al-Amid Chronique des Ayyoubides 602–658/1205–1260* (Paris, 1994).

Elad, Amikam. "The Ethnic Composition of the Abbasid Revolution: A Re-evaluation of Some Recent Research", *JSAI* 24 (2000): 275–82.

El Cheikh, Nadia Maria. "Revisiting the Abbasid Harems", *Journal of Middle East Women's Studies* 1/3 (2005): 1–19.

Eph'al, Israel. *The Ancient Arabs: Nomads on the Borders of the Fertile Crescent 9th-5th Centuries B.C.* (Jerusalem, 1982).

Esin, Emel. "Tarkhan Nizak or Tarkhan Tirek? Enquiry Concerning the Prince of Badhghis Who in A.H. 9 1 /709–710 Opposed the 'Omayyad Conquest of Central Asia'", *JAOS* 97 (1977): 323–31.

Fackenheim, Emil L. "The Possibility of the Universe in Al-Farabi, Ibn Sina and Maimonides", *Proceedings of the American Academy for Jewish Research* 16 (1946–47): 39–70.

Fakhry, Majid. "The Platonism of Miskawayh and Its Implications for His Ethics", *SI* 42 (1975): 39–57.

Findley, Carter V. *The Turks in World History* (Oxford, 2005)

Forand, Paul. "The Governors of Mosul According to al-Azdī's Ta'rīkh al-mawṣil", *JAOS* 89/1 (1969): 88–105.

Frenkel, Y. "The Turks of the Eurasian Steppes in Arabic Geographical Literature", in M. Biran and R. Amitai (eds.), *Mongols, Turks and Others: Eurasian Nomads and the Sedentary World* (Leiden, 2005), 201–41.

Friedmann, Y. "Finality of Prophethood in Sunni Islam", *JSAI* 7 (1986): 177–215.

Frye, R.N. "The Samanids: A Little-Known Dynasty", *The Muslim World* 34 (1944): 40–5.

—. "A New Arabic Geographical Manuscript", *JNES* 8 (1949): 90–7.

—. *Bukhara: The Medieval Achievement* (Norman, OK, 1965).

—. *The Golden Age of Persia* (London, 1989).

— (trans.). *Ibn Fadlan's Journey To Russia: A Tenth Century Traveler From Baghdad to the Volga River* (Princeton, 2005).

— and Robert Blake. "History of the Nation of the Archers (The Mongols) by Grigor of Akanc; Hitherto Ascribed to Magakia the Monk: The Armenian Text edited with an English Translation and Notes", *Harvard Journal of Asiatic Studies* 12 (1949): 269–443.

— and Aydin M. Sayili. "Turks in the Middle East before the Saljūqs", *JAOS* 63 (1943): 194–207.

—, with R. Blake. "Notes on the Risala of Ibn Fadlan", *Bizantina Metabyzantina* 1 (1949): 7–37.

Gandjeï, Tourkhan. "Turkish in Pre-Mongol Persian Poetry", *Bulletin of the School of Oriental and African Studies* 49 (1986): 67–75.

Gardet, Louis. "Les noms et les statuts. Le problème de la foi et des œuvres en Islām", *SI* 5 (1956): 61–123.

van Gelder, Geert Jan. "Mirror for Princes or Vizor for Viziers: The Twelfth-Century Arabic Popular Encyclopedia '*Mufid al-ʿulūm*' and Its Relationship with the Anonymous Persian '*Bahr al-fawa'id*'", *BSOAS* 64 (2001): 313–38.

Ghalib, Ismaʿil. *Maskukat qadima islamiyya qataluki* (Istanbul, 1894).

Gibb, Hamilton A.R. "Kleine Mitteilungen und Anzeigen Zu Der Islam XIV 402—406", *Der Islam* 20 (1932): 196–7.

—. "The Social Significance of the Shu'ubiya", *Studies on the Civilization of Islam* (London and Boston, 1962), 62–73.

—. "Al-Mawardi's Theory of the Caliphate", in his *Studies on the Civilization of Islam* (London and Boston, 1962), 151–65.

Goitein, S.D. "The Origin of the Vizierate and its True Character", *Islamic Culture* 16 (1942): 255–62, 380–92.

—. "On the Origin of the Term Vizier", *JAOS* 81/4 (1961): 425–6.

Golden, P.B. "The Migrations of the Oguz", *Archivim Ottomanicum* 4 (1972): 45–84. [repr. in his *Nomads and their Neighbours in the Russian Steppe* (London, 2003)].

—. "Wolves, Dogs and Qipcaq religion", *Acta Orientalia* 50 (1977): 87–97 [repr. in his *Nomads and their Neighbours in the Russian Steppe* (London, 2003)].

—. *Khazar Studies: An Historico-Philological Inquiry into the Origins of the Khazars* (Budapest, 1980).

—. "The Peoples of the South Russian Steppes", in Denis Sinor (ed.), *The Cambridge History of Early Inner Asia* (Cambridge, 1990), 256–84.

— "'I Will Give the People unto Thee': The Činggisid Conquests and Their Aftermath in the Turkic World", *Journal of the Royal Asiatic Society* 10/1 (2000): 21–41.

—. "The Qipcaqs of Medieval Eurasia: An Example of Stateless Adaptation in the Steppes", in G. Seaman and D. Marks (eds.), *Rulers from the Steppe: State Formation on the Eurasian Periphery* (Los Angeles 1991) [repr. in his *Nomads and their Neighbours in the Russian Steppe* (London, 2003)].

—. *An Introduction to the History of Turkic Peoples: Ethnogenesis and State-Formation in Medieval and Early Modern Eurasia and the Middle East* (Wiesbaden, 1992).

Goldziher, Ignaz. *Muslim Studies*, ed. and trans. S.M. Stern (London, 1967).

Gordon, Matthew S. "The Khaqanid Families of the Early ʿAbbasid Period", *Journal of the American Oriental Society* 121 (2001): 236–55.

—. *The Breaking of a Thousand Swords: A History of the Turkish Community of Samarra (200–275/815–889)* (Albany, NY, 2001).

Grabar, O. "Paintings of the Six Kings at Qusayr 'Amrah", *Ars Orientalis* 1 (1955): 185–7.

Graf, Gunhild. *Die Epitome der Universalchronik Ibn ad-Dawadaris im Verhaltnis zur Langfassung: eine quellenkritische Studie zur Geschichte der agyptischen Mamluken* (Berlin, 1990).

Grignaschi, M. "Les *rasail Aristatalisa ila Iskandar* de Salim Abū-l-ʿAla et l'activite culturelle a l'epoque Omayyade", *Bulletin d'Etudes Orientales* 19 (Damascus, 1965): 7–83.

—. "Le Roman Epistolaire classique conserve dans la version Arabe de Salim Abū al-ʿAla", *Le Muséon* 80 (1967): 211–64.

—. "La Nihayatu-l-ʿArab fi ahbari-l-Furs wa-l-ʿArab (premiere partie)", *Bulletin des Études Orientales* 22 (1969): 17–67.

Gronke, Monika. "The Arabic Yarkand Documents", *BSOAS* 49 (1986): 454–507.

Guo, Li. "Mamluk Historiographic Studies: The State of the Art", *Mamluk Studies Review* 1 (1997): 15–43.

—. "Paradise Lost: Ibn Daniyal's Response to Sultan Baybars' Campaign against Vice in Cairo", *JAOS* 121/2 (2001): 219–35.

Haarmann, Ulrich. "Altun Han und Cingiz Han bei den aegyptischen Mamluken", *Der Islam* 51/1 (1974): 16–23.

Hachmeier, Klaus S. "Private Letters, Official Correspondence: Buyid *inshā'* as a Historical Source", *Journal of Islamic Studies* 13 (2002): 125–54.

Hajib, Yusuf Khass. *Wisdom of Royal Glory: (Kutadgu Bilig) A Turko-Islamic Mirror for Princes,* trans. R. Dankoff (Chicago, 1983).

Hallaq, Wael B. "Caliphs, Jurists and the Saljūqs in the Political Thought of Juwayni", *The Muslim World* 74 (1984): 26–41.

Halm, Heinz. *Shiism* (Edinburgh, 1991).

Hamada, M. "Le Mausolee et le cults de Satuq Bughara Khan", *Journal of the History of Sufism* 3 (2001): 63–87.

Hamdani, Viqar Ahmed. "Some Rare Manuscripts in Istanbul", *JRAS* (1938), 561–64.

Hasson, Isaac. *Recherches sur Muawiya Ibn Abi Sufiyan*, Ph.D. dissertation, The Hebrew University of Jerusalem, 1982; Hebrew.

—. "Judham entre la Jahiliyya et l'Islam", *Studia Islamica* 81 (1995): 5–41.

Herman, Geoffrey. "Ahasuerus, the Former Stable-Master of Belshazzar, and the Wicked Alexander of Macedon: Two Parallels Between the Babylonian Talmud and Persian Sources", *AJS Review* 29:2 (2005): 283–97.

Herzfeld, E. "Die Könige der Erde", *Der Islam*, 21 (1933): 233–6.

Hillenbrand, Carole. *Turkish Myth and Muslim Symbol: The Battle of Manzikert* (Edinburgh, 2007).

Hillenbrand, R. *Islamic Architecture* (New York, 1994).

Hinds, Martin. "The Murder of the Caliph Uthman", *IJMES* 3 (1972): 450–69.

Hodgson, Marshall G. *The Venture of Islam: Conscience and History in a World Civilization* vol. 2: *The Expansion of Islam in the Middle Periods* (Chicago, 1974).

Holt, Peter Malcolm. "The Exalted Lineage of Riḍwan Bey: Some Observations on a Seventeenth-Century Mamluk Genealogy", *BSOAS* 22 (1959): 221–30.

—. "The Sultanate of al-Mansur Lachin (696–8/1296–9)", *BSOAS* 36 (1973): 521–32.

Hourani, George F. "Ghazālī on the Ethics of Action", *JAOS* 96/1 (1976): 69–88.

Houtsma, M.T. *Ein türkisch-arabisches Glossar nach der Leidener Handschrift* (Leiden, 1894).

Hutchins, W.M. *Nine Essays of al-Jāḥiẓ* (New York, 1975).

Ibish, Yusuf. *The Political Doctrine of Baqillani* (Beirut, 1966).

Inostranzev, Konstantin A. *Iranian Influence on Moslem Literature [part 1 translated with supplementary appendices from Arabic sources by G. K. Nariman]* (Bombay, 1918).

Irwin, Robert. "Mamluk History and Historians", in Roger Allen and D. S. Richards (eds.), *The Cambridge History of Arabic Literature: Arabic Literature in the Post-Classical Period* (Cambridge, 2008), 159–70.

Ismail, Osman S.A. "Mu'tasim and the Turks", *Bulletin of the School of Oriental and African Studies* 29 (1966): 12–24.

—. "The Founding of a New Capital: Samarra", *Bulletin of the School of Oriental and African Studies* 31 (1968): 1–13.

Jacob, Xavier (trad.). *Les Turcs au Moyen-Age* (Ankara, 1990).

Johansen, Babar. *Contingency in a Sacred Law: Legal and Ethical Norms in the Muslim Fiqh* (Leiden, 1999).

Jones, J.M.B. "The Chronology of the "Maghāzī" – A Textual Survey", *BSOAS* 19/2 (1957): 245–80.

Kabir, M. *The Buwayhid Dynasty of Baghdad* (Calcutta, 1964).

Kafesoğlu, Ibrahim. "A Propos du nom Türkmen", *Oriens* 11 (1958): 146–50.

Kahle, Paul. "Islamische Quellen zum chinesischen Porzellan", *ZDMG* 88 (1934): 12–45.

Kennedy, Hugh. "Central Government and Provincial Elites in the Early ʿAbbāsid caliphate", *BSOAS* 44 (1981): 26–38.

Khalidi, Tarif. *Arabic Historical Thought in the Classical Period* (Cambridge, 1994).

Khalidov, A.B. "Myth and Reality in the *Kitāb akhbār al-buldān* by Ibn al-Faqīh", in Angelika Neuwirth et al. (eds.), *Myths, Historical Archetypes and Symbolic Figures in Arabic Literature: Towards a New Hermeneutic Approach* (Stuttgart, 1999), 481–9.

Khan, Geoffrey. "The Historical Development of the Structure of Medieval Arabic Petitions", *BSOAS* 53 (1990): 8–30.

Khan, M.S. "A Manuscript of an Epitome of al-Ṣābi's *Kitāb al-Tajī*", *Arabica* 12 (1965): 27–44.

—. "The Contents of the *Kitāb al-Taji* A Manuscript of Abū Isḥāq Ibrahim al-Ṣābi", *Islamic Studies* 8 (1968): 247–52.

—. "Miskawaih and Arabic Historiography", *Journal of the American Oriental Society* 89 (1969): 710–30.

—. "Studies in the *Kitāb al-Taji*, of al-Ṣābi", *Arabica* 17 (1970): 151–60.

Klausner, Carla L. *The Seljuk Vezirate: A Study of Civil Administration 1055–1094* (Cambridge, MA, 1973).

Köprülü, Mehmet Fuat. *The Origins of the Ottoman Empire*, trans. Gary Leiser (Albany, NY, 1991).

—. *Early Mystics in Turkish Literature*, trans. and ed. Gary Leiser and Robert Dankoff (London, 2006).

Kouymjian, Dickran. "Problems of Medieval Armenian and Muslim Historiography: The Mxit'ar of Ani Fragment", *IJMES* 4 (1973): 465–75.

Kowalski, T. "Die aeltesten Erwaehnungen der Tuerken in der arabischen Literatur", *Koeroesi Csoma-Archivum*, no. 2 (Budapest, 1932), 38–41.

Kraemer, Joel L. "Humanism in the Renaissance of Islam: A Preliminary Study", *Journal of the American Oriental Society* 104 (1984), 135–64.

Kucur, Sadi S. "A Study on the Coins of Ṭughril Beg, the Sultan of the Great Seljuqs", in C. Alfaro, C. Marcos and P. Otero (eds.), *Actas del XIII Congreso Internacional de Numismática* (Madrid, 2003/2005), 2: 1599–608.

van Laer, Z. "Ibn al-Faqīh et le Livre des pays", *Arabica* 31 (1984): 69–79.

Lagrange, Frédéric. "L'Obscénité du Vizir", *Arabica* 53/1 (2006): 54–107.

Lambton, Ann K.S. "Justice in the Medieval Persian Theory of Kingship", *SI* 17 (1962): 91–119.

—. *State and Government in Medieval Islam: An Introduction to the Study of Islamic Political Theory: The Jurists* (Oxford, 1981).

—. "Changing Concepts of Justice and Injustice from the 5th/11th Century to the 8th/14th Century in Persia: The Saljūq Empire and the Ilkhanate", *SI* 68 (1988): 27–60.

Landau-Tasseron, Ella. "The 'Cyclical Reform': A Study of the Mujaddid Tradition", *Studia Islamica* 70 (1989): 79–117.

Lane-Poole, S. *Catalogue of Oriental Coins in the British Museum* (London, 1875–90).

—. *Additions to the Oriental Collection* (London, 1889).

Lapidus, I.M. *A History of Islamic Societies* (Cambridge, 1988).

Lassner, Jacob. "Why Did the Caliph Al-Manṣūr Build Ar-Ruṣafah? A Historical Note", *JNES* 24 (1965): 95–9.

—. "Abū Muslim al-Khurāsānī: The Emergence of a Secret Agent from Khurāsān, Irāq, or Was it Iṣfahān?", *JAOS* 104/1 (1984): 165–75.

Lecomte, Gérard. "Le problème d'Abū ʿUbayd: Réflexions sur les 'erreurs' que lui attribue Ibn Qutayba", *Arabica* 12/2 (1965): 140–74.

Leder, Stefan. *Ibn al-Gauzi und seine Kompilation Wider die Leidenscaft* (Beirut, 1984).

— and Hilary Kilpatrick, "Classical Arabic Prose Literature: A Researcher's Sketch Map", *Journal of Arabic Literature* 23/1 (1992): 2–26.

Lev, Y. *State and Society in Fatimid Egypt* (Leiden, 1991).

Levi, S.C. and R. Sela (eds.) *Islamic Central Asia: An Anthology of Historical Sources* (Bloomington, 2010).

Lewicki, T. *Źródła arabskie do dziejów słowiańszczyzny* (Wrocław, 1969).

—. "Turkische Stadt Dana nach einem Bericht des arabischen Geographen Ibn al-Faqih", *Folia Orientalia* 13 (1971): 161–70.

Lewis, B. *The Assassins* (London, 1967).

—. *Race and Colour in Islam* (New York, 1974).

—. (ed. and trans.), *Islam from the Prophet Muhammad to the Capture of Constantinople*, 2 vols. (New York, 1974).

—. *Race and Slavery in the Middle East: A Historical Enquiry* (New York, 1990).

MacDonald, Duncan B. "The Earlier History of the Arabian Nights", *JRAS* (1924): 353–97.

Madelung, W. "Abū Isḥāq al-Ṣābi on the ʿAlīds of Tabaristan and Gilan", *JNES* 26 (1967), 17–57.

—. "The Assumption of the Title Shahanshah by the Buyids and the Reign of the Daylam", *JNES* 28 (1969): 84–108, 168–83.

—. "Notes on al-Ṣābi's Kitāb al-Taji", *Islamic Studies* 9 (1970), 81–8.

—. "The Origin of the Controversy concerning the Creation of the Koran", *Orientalia Hispanica* 1 (1974): 504–25 [repr. in his *Religious Schools and Sects* (Variorum, 1985), article 5].

—. *The Succession to Muhammad: A Study of the Early Caliphate* (Cambridge, 1997).

Makdisi, George. "Autograph Diary of an Eleventh Century Historian of Baghdad", *BSOAS* 18 (1956): 9–31, 239–60; 19 (1957) 13–48, 281–303, 426–43.

—. "The Marriage of Tughril Beg", *IJMES* 1 (1970): 259–75 [repr.in his *History and Politics in Eleventh Century Baghdad* (London, 1990), art. 9].

—. "Les Rapports Entre Calife et Sultan a l'epoque Saljūqide", *IJMES* 6/2 (1975): 228–36.

Manzalaoui, Mahmoud. "The Pseudo-Aristotelian '*Kitab Sirr al-asrar*': Facts and Problem", *Oriens* 23 (1974): 147–257.

Maroth, M. "The Correspondence between Aristotle and Aexander the Great: An Anonymous Greek Novel in Letters in Arabic Translation", *Acta Antiqua Hungaricae* 45 (2005): 231–315.

Marquart, J. "Beiträge zur Geschichte und Sage von Erän", *Zeitschrift der Deutschen Morgenländischen Gesellschaft* 49 (1895): 628–72.

—. *Osteuropäische und ostasiatische Streifzüge Ethnologische und historisch-topographische Studien zur Geschichte des 9. und 10. Jahrhunderts (ca. 840–940)* (Leipzig, 1903).

Mason, Kenneth. "Notes on the Canal System and Ancient Sites of Babylonia in the Time of Xenophon", *The Geographical Journal* 56/6 (1920): 468–81.

Meisami, Julie Scott. *Persian Historiography to the End of the Twelfth Century* (Edinburgh, 1999).

Melville, Charles. "History: From the Saljūqs to the Aq Qoyunlu (ca. 1000–1500 C.E.)", *Iranian Studies* 31 (1998): 473–82.

Meyerhof, Max. "An Arabic Compendium of Medico-Philosophical Definitions", *Isis* 10 (1928): 340–9.

—. "'Alī at-Tabari's 'Paradise of Wisdom', One of the Oldest Arabic Compendiums of Medicine", *Isis* 16 (1931): 7–11.

Miles, G.C. *The Numismatic History of Rayy* (New York, 1938).

Minorsky, V. "La Domination des Dailamites" (Paris, 1932) [repr. in his *Iranica* (Tehran, 1964), 12–30].

—. "A Civil and Military Review in Fars in 881/1476", *BSOAS* 1 (1939): 141–78.

— (ed. and trans.). *Marvazi on China, the Turks and India: Arabic text (c. A.D. 1120) with an English translation and commentary* (London: The Royal Asiatic Society, 1942).

—. "Tamim ibn Bahr's Journey to the Uyghurs", *BSOAS* 12 (1948): 275–305 [repr. in his *The Turks, Iran, and the Caucasus in the Middle Ages* (London, 1978)].

—. "La deuxieme risāla d'Abū-Dulaf", *Oriens* 5/1 (1952): 23–7.

—. "Caucasica IV", *BSOAS* 15/3 (1953): 504–29.

Miquel, André. *La Geographie humaine dumonde musulman jusqu'au milieu du 11e siecle 2– Geographie arabe et represetation du monde: la terre et l'etranger* (Paris, 1975).

Montgomery, James. "Ibn Fadlan and the Rusiyyah", *Journal of Arabic and Islamic Studies* 3 (2000): 1–25.

Mottahedeh, Roy. "The Shu`ubiyah Controversy and the Social History of Early Islamic Iran", *IJMES* 7 (1976): 161–82.

Murgotten, Francis Clark (trans.). *The Origins of the Islamic State: Being a Translation from the Arabic, Accompanied with Annotations, Geographic and Historic Notes of the Kitab futuh al-buldan* (New York, 1924).

Nagel, T. *Alexander der Grosse in der fruhislamischen Volksliteratur* (Walldorf-Hessen, 1978).

Nagrodzka-Majchrzyk, T. "Les Oghouz dans la relation d'Aḥmad Ibn Fadlan", *Rocznik orientalistyczny* 49 (1995): 165–9.

Nazim, Muhammad. *The Life and Times of Sultan Mahmud of Ghazna* (New Delhi, 1931; 2nd edn, 1971).

Nazmi, A. "The Functions of the White Slaves in the Social Life of the Abbasid Caliphate", *Rocznik Orientalistyczny* 51 (1998): 91–103.

Netton, I.R. *al-Farabi and his School* (London, 1992).

Newby, G. *The Making of the Last Prophet: A Reconstruction of the Earliest Biography of Muḥammad* (Columbia, SC, 1999).

Noldeke, T. "Beitrage zur Geschichte des Alexander-Romans", *Osterreichische Akademie der Wissenschaften, Denkschriften der Philosophisch Historische Klasse* CL. 38. Bd.5. Abh (Wien, 1890), 34–50.

—. *The Iranian National Epic or the Shahnamah*, trans. L.T. Bogdanov (Bombay, 1930; repr. Philadelphia, 1979).

Norris, H.T. "Shu'ubiyya in Arabic Literature", in Julia Ashtiany et al. (eds.), *The Cambridge History of Arab Literature: Abbasid Belles-Lettres* (Cambridge, 1990), 31–47.

Obermann, Julian. "Political Theology in Early Islam: Ḥasan Al-Baṣri's Treatise on Qadar", *JAOS* 55 (1935): 138–62.

Ouseley, William. *The Oriental Geography of Ebn Haukal, An Arabian Traveller of the Tenth Century* (London, 1800).

Peacock, Andrew C.S. *Early Seljūq History* (London, 2010).

Pellat, C. "Ğāhiẓiana, I: Le *Kitāb al-Tabaṣṣur bi-l-tiğāra* attribué à Ğāḥiẓ", *Arabica* 1/2 (1954): 153–65.

Pessagno, J. Meric. "The Uses of Evil in Maturidian Thought", *SI* 60 (1984): 59–82.

Peters, Rudolph. *Jihad in Mediaeval and Modern Islam* (Leiden, 1977).

Pipes, Daniel. *Slave Soldiers and Islam: The Genesis of a Military System* (New Haven, CT, 1981).

Pourshariati, Parvaneh, "Local Historiography in Early Medieval Iran and the Tārīkh-i Bayhaq", *Iranian Studies* 33 (2000): 133–64.

Pritsak, O. "Von den Karluk zu den Karachaniden", *ZDMG* 101 (1951): 270–300.

—. "Die Karachaniden", *Der Islam* 31 (1953): 17–63.

Pulleyblank, Edwin G. "A Sogdian Colony in Inner Mongolia", *T'oung Pao* Second Series 41–4/5 (1952): 317–56.

al-Qaddumi, Ghada al-Hijjawi (trans.). *The Book of Gifts and Rarieties: Selections Compiled in the Fifteenth Century from an Eleventh-Century Manuscript on Gifts and Treasures* (Cambridge, MA, 1996).

Rabbat, Nasser O. "The Ideological Significance of the Dar al-ʿAdl in the Medieval Islamic Orient", *IJMES* 27 (1995): 3–28.

Rabino, H.L. *Mazandaran and Astarabad* (London, 1928).

Rāghib, Yousef and Philippe Fluzin. "La fabrication des lames damassées en Orient", *JESHO* 40/1 (1997): 30–72.

Reynolds, Gabriel Said. "A Reflection on Two Qurʾānic Words (Iblīs and Jūdī), with Attention to the Theories of A. Mingana", *JAOS* 124/4 (2004): 675–89.

Rice, D.S. "Medieval Harran: Studies on Its Topography and Monuments, I", *Anatolian Studies* 2 (1952): 36–84.

Richter-Bernburg, Lutz. "Linguistic Shuʿūbīya and Early Neo-Persian Prose", *JAOS* 94/1 (1974): 55–64.

—. "Amir-Malik-Shahanshah: ʿAḍud ad-Daula's Titulature Re-examined", *Iran* 18 (1980): 83–102.

Riefstahl, Rudolf J. "Vier Syrische Marmorkapitale mit figuralen Darstellungen in der Moschee zu Boz Uyuk", *Der Islam* 20 (1932): 186–95.

Rodinson, Maxime. *Mohammed* (Harmondsworth, 1971).

Rogers, J.M. "Waqf and Patronage in Seljuk Anatolia: The Epigraphic Evidence", *Anatolian Studies* 26 (1976): 69–103.

von Rohr-Sauer, A. *Abū Dulaf Bericht uber seine Reise nach Turkestan* (Bonn, 1933).

Rosenthal, Erwin I.J. *Political Thought in Medieval Islam: An Introductory Outline* (Cambridge, 1962).

Rosenthal, Franz. "Ar-Razi on the Hidden Illness", *Bulletin of the History of Medicine* 43 (1969): 45–60.

—. "'Blurbs' (*taqrīẓ*) from Fourteenth-Century Egypt", *Oriens* 27 (1981): 177–96.

Roux, Jean-Paul. "Mosquées anatoliennes à décor figuratif sculpté", *Syria* 57/1 (1980): 305–23.

Rowson, Everett K. "The Effeminates of Early Medina", *JAOS* 111 (1991): 673–91.

Roxburgh, David J. "On the Transmission and Reconstruction of Arabic Calligraphy: Ibn al-Bawwab and History", *SI* 96 (2003): 39–53.

Rypka, Jan. *A History of Iranian Literature* (Dordrecht, 1968).

Sadan, J. (ed.). *Nouvelle source sur l'epoque Buyide* (Tel-Aviv, 1980).

Safi, Omid. *The Politics of Knowledge in Pre-modern Islam: Negotiating Ideology and Religious Inquiry* (Chapel Hill, NC, 2006).

al-Saqa, M. (ed.). *Mukhtār al-shiʿr al-jāhilī* (Cairo, 1969).

Savvides, A.G.C. "Some Notes on the Terms Khan and Khagan in Byzantine Sources", in I.R. Netton (ed.), *Studies in Honour of Clifford Edmond Bosworth* (Leiden, 2000), 1: 267–79.

Schacht, Joseph. *The Origins of Muhammadan Jurisprudence* (Oxford, 1950/1967).

von Schlözer, C. *Abū Dolef Misaris de itinere Asiatico commentarius* (Berlin, 1845).

Schönig, C. "On Some Unclear, Doubtful and Contradictory Passages in Maḥmūd al-Kāšġari's 'Dīwān Luġāt at-Turk'", *Türk Dilleri Araştırmaları* 14 (2004): 37–8.

Senga, T. "The Toquz Oghuz Problem and the Origin of the Khazars", *Journal of Asian History* 24/1 (1990): 57–61.

Serjeant, R.J. "Material for a History of Islamic Textiles up to the Mongol Conquest", *Ars Islamica* 15–16 (1951): 29–85.

—. "Star-Calendars and an Almanac from South-West Arabia", *Anthropos* 49/3–4 (1954): 478–502.

Sesen, R. *Hilafet ordusunun menkýbeleri ve Türklerin fazailetleri* (Ankara, 1967).

Sezgin, F. (ed.) *Qudama ibn Ja'far- Book on Taxation and Official Correspondence* (Frankfurt, 1986).

—. (compiler). *Islamic Geography 169: Studies on the Travel Account of Ibn Fadlan and Abū Dulaf (first half 10th century)* (Frankfurt am Main: Institute for the History of Arabic-Islamic Science, 1994).

—. (compiler). *Islamic Geography 184: Studies on Abū Hamid al-Garnati* (Frankfurt am Main: Institute for the History of Arabic-Islamic Science, 1994).

Shahid, Irfan. *Byzantium and the Arabs in the Sixth Century* (Washington, DC, 2002).

Shaked, S. "From Iran to Islam: On Some Symbols of Royalty", *JSAI* 7 (1986): 75–91 [repr. in his *From Zoroastrian Iran to Islam: Studies in Religious History and Intercultural Contacts* (Aldershot, 1995) art. 7].

Sharon, M. "Notes on the Question of the Legitimacy of Government in Islam", *Israel Oriental Studies* 10 (1980): 116–23.

—. "The Development of the Debate around the Legitimacy of Authority in Early Islam", *JSAI* 5 (1984): 121–42.

—. "*Ahl al-Bayt* – The People of the House", *JSAI* 8 (1986): 174–80.

—. *Revolt: The Social and Military Aspects of the ʿAbbasid Revolution* (Jerusalem, 1990).

—. "The Umayyads as *Ahl al-Bayt*", *JSAI* 14 (1991): 135–43.

Silverstein, Adam J. *Postal Systems in the Pre-modern Islamic World* (Cambridge, 2007).

Smith, John Masson Jr. "Turanian Nomadism and Iranian Politics", *Iranian Studies* 11 (1978): 57–81.

Soucek, Svat. *A History of Inner Asia* (Cambridge, 2000).

Southgate, Minoo S. "Portrait of Alexander in Persian Alexander-Romances of the Islamic Era", *JAOS* 97/3 (1977): 278–84.

Sprenger, Aloys. *Die Post- und Reiserouten des Orients* (Leipzig, 1864).

Stern, S. M. "The Arabic Translation of the Pseudo-Aristotelian Treatise *De Mundo*", *Le Museon* 77 (1964): 187–204.

—. "A Third Arabic Translation of the Pseudo-Aristotelian Treatise *De Mundo*", *Le Museon* 78 (1965): 381–93.

—. *Aristotle on the World-State* (Oxford, 1968).

—. "Yaʿqūb the Coppersmith and Persian National Sentiment", in C.E. Bosworth (ed.), *Iran and Islam: Studies in Memory of the late V. Minorsky* (Edinburgh, 1971), 535–55.

Stillman, Yedida K. and Norman A. Stillman. *Arab Dress: A Short History from the Dawn of Islam to Modern Times* (Leiden, 2007).

Le Strange, G. "Description of Mesopotamia and Baghdād, Written about the Year 900 A.D. by Ibn Serapion. The Arabic Text Edited from a MS. in the British Museum Library, with Translation and Notes", *JRAS* (1895): 1–76, 255–315.

—. "Baghdad During the Abbasid Caliphate. A Topographical Summary with a Note of the Contemporary Arabic and Persian Authorities", *JRAS* (1899): 847–93.

—. *The Lands of the Eastern Caliphate* (London, 1905).

Sublet, Jacqueline. "Nom écrit, nom dit: Les personnages du théâtre d'ombres d'Ibn Dāniyāl", *Arabica* 44/4 (1997): 545–52.

Suemer, Faruk, Ahmet E. Uysal and, Warren S. Walker (trans.). *The Book of Dede Korkut: A Turkish Epic* (Austin, 1972).

Tavakoli-Targhi, Mohamad. "Contested Memories: Narrative Structures and Allegorical Meanings of Iran's Pre-Islamic History", *Iranian Studies* 29 (1996): 149–75.

Tezcan, Baki. "Hanafism and the Turks in al-Ṭarasūsī's Gift for the Turks (1352)", *Mamluk Studies Review* 15 (2011): 67–86.

Tillier, Mathieu. "L'Exemplarité chez al-Tanùkhì: les cadis dans le *Nishwàr al-Muhadara", Arabica* 44 (2007): 1–24.

Thatcher, G.W. *Arabic Grammar of the Written Language* (New York, 2003).

Tor, D.G. "The Mamluks in the Military of the Pre-Seljuq Persianate Dynasties", *Iran* 46 (2008): 213–25.

Treadwell, Luke. "The Account of the Samanid Dynasty in Ibn Ẓāfir al-Azdī's *Akhbār al-duwal al-munqaṭi'a*", *Iran* 43 (2005): 135–71.

Tskitishvili, O. "Concerning the Tashkent manuscript of a geographical treatise by Ibn al-Faqīh al-Hamadani", *JESHO* 24 (1980): 304–8.

Tsuen-Hsuin, Tsien. *Chemistry and Chemical Technology* pt. 1 *Paper and Printing* (Cambridge, 1985).

De Vaux, Carré. "Notice sur un calendrier turc", in *A Volume of Oriental Studies presented to G. Browne* (Cambridge, 1922), 106–16.

Wagner, Ewald. *Abū Nuwas: eine Studie zur arabischen Literatur der fruhen 'Abbāsidenzeit* (Wiesbaden, 1965).

Waldman, Marilyn R. *Toward a Theory of Historical Narrative: A Case Study in Perso-Islamicate Historiography* (Columbus, OH, 1980).

Watt, W. Montgomery. *Muhammad at Mecca* (Oxford, 1953).

—. *Islamic Creeds: A Selection* (Edinburgh, 1994).

Weigert, Gideon. "A Note on Hudna: Peace Making in Islam", in Y. Lev (ed.), *War and Society in the Eastern Mediterranean, 7–15th Centuries* (Leiden, 1997), 399–406.

Weintritt, Otfried. *Formen spatmittelalterlicher islamischer Geschichtsdarstellung: Untersuchungen zu an-Nuwairi al-Iskandaranis Kitab al-Ilmam und verwandten zeitgenossischen Texten* (Beirut, 1992).

White, Hayden. "The Value of Narrativity in the Representation of Reality", in W.J.T. Mitchell (ed.), *On Narrative* (Chicago and London, 1981), 1–23.

Wiet, G. *L'Expedition Persane* (Cairo, 1931).

Williams, Steven J. *The Secret of Secrets: The Scholarly Career of a Pseudo-Aristotelian text in the Latin Middle Ages* (Ann Arbor, MI, 2003).

Wilson, C.E. "The Wall of Alexander against Gog and Magog and the Expedition sent out to find it by the Khalif al-Wathiq in 842 A.D.", in *Asia Major. Hirth Anniversary Volume* (London, 1923), 575–612.

Wittek, P. "Eine Turkische Furstin auf dem Wandgemalde von Kusajr 'Amra", *Der Islam* 14 (1924): 402–6.

Wüstenfeld, F. "Des Abū Dulaf Bericht über die türkischen Horden", *Zeitschrift für vergleichende Erdkunde* 2 (1842): 205 ff.

Yarshater, Ehsan. "Iranian National History", in *The Cambridge History of Iran* (Cambridge, 1983), 3: 359–478.

Yavari, Neguin. "Mirrors for Princes or a Hall of Mirrors? Niẓām al-Mulk's Siyar al-mulūk Reconsidered", *al-Masaq* 20/1 (2008): 47–69.

Zakariya Kitabji (Zekeriya Kýtapchý). *al-Turk fī mu'allifāt al-Jāḥiẓ wa-makānātahum fī al-Ta'rīkh al-islāmī ḥattā al-qarn al-thālith al-hijrī* [The Turks in the Works of al-Jāḥiẓ and their Position in Islamic History] (Beirut, 1972).

Zakeri, Mohsen. *Sasanid Soldiers in Early Muslim Society: The Origin of 'Ayyaran and Futuwwa* (Wiesbaden, 1995).

Index